Surveying Victims:
Options for Conducting the National Crime Victimization Survey

Panel to Review the Programs of the Bureau of Justice Statistics

Robert M. Groves and Daniel L. Cork, *Editors*

Committee on National Statistics
Committee on Law and Justice

Division of Behavioral and Social Sciences and Education

NATIONAL RESEARCH COUNCIL
OF THE NATIONAL ACADEMIES

THE NATIONAL ACADEMIES PRESS
Washington, DC
www.nap.edu

THE NATIONAL ACADEMIES PRESS   500 Fifth Street, NW   Washington, DC 20001

NOTICE: The project that is the subject of this report was approved by the Governing Board of the National Research Council, whose members are drawn from the councils of the National Academy of Sciences, the National Academy of Engineering, and the Institute of Medicine. The members of the committee responsible for the report were chosen for their special competences and with regard for appropriate balance.

The project that is the subject of this report was supported by contract no. SES-0453930 between the National Academy of Sciences and the U.S. Department of Justice. Support of the work of the Committee on National Statistics is provided by a consortium of federal agencies through a grant from the National Science Foundation (Number SBR-0112521). Any opinions, findings, conclusions, or recommendations expressed in this publication are those of the author(s) and do not necessarily reflect the views of the organizations or agencies that provided support for the project.

International Standard Book Number-13: 978-0-309-11598-8
International Standard Book Number-10: 0-309-11598-1

Additional copies of this report are available from the National Academies Press, 500 Fifth Street, NW, Washington, DC 20001; (202) 334-3096; Internet, http://www.nap.edu.

Suggested citation: National Research Council (2008). *Surveying Victims: Options for Conducting the National Crime Victimization Survey.* Panel to Review the Programs of the Bureau of Justice Statistics. Robert M. Groves and Daniel L. Cork, eds. Committee on National Statistics and Committee on Law and Justice, Division of Behavioral and Social Sciences and Education. Washington, DC: The National Academies Press.

THE NATIONAL ACADEMIES
Advisers to the Nation on Science, Engineering, and Medicine

The **National Academy of Sciences** is a private, nonprofit, self-perpetuating society of distinguished scholars engaged in scientific and engineering research, dedicated to the furtherance of science and technology and to their use for the general welfare. Upon the authority of the charter granted to it by the Congress in 1863, the Academy has a mandate that requires it to advise the federal government on scientific and technical matters. Dr. Ralph J. Cicerone is president of the National Academy of Sciences.

The **National Academy of Engineering** was established in 1964, under the charter of the National Academy of Sciences, as a parallel organization of outstanding engineers. It is autonomous in its administration and in the selection of its members, sharing with the National Academy of Sciences the responsibility for advising the federal government. The National Academy of Engineering also sponsors engineering programs aimed at meeting national needs, encourages education and research, and recognizes the superior achievements of engineers. Dr. Charles M. Vest is president of the National Academy of Engineering.

The **Institute of Medicine** was established in 1970 by the National Academy of Sciences to secure the services of eminent members of appropriate professions in the examination of policy matters pertaining to the health of the public. The Institute acts under the responsibility given to the National Academy of Sciences by its congressional charter to be an adviser to the federal government and, upon its own initiative, to identify issues of medical care, research, and education. Dr. Harvey V. Fineberg is president of the Institute of Medicine.

The **National Research Council** was organized by the National Academy of Sciences in 1916 to associate the broad community of science and technology with the Academy's purposes of furthering knowledge and advising the federal government. Functioning in accordance with general policies determined by the Academy, the Council has become the principal operating agency of both the National Academy of Sciences and the National Academy of Engineering in providing services to the government, the public, and the scientific and engineering communities. The Council is administered jointly by both Academies and the Institute of Medicine. Dr. Ralph J. Cicerone and Dr. Charles M. Vest are chair and vice chair, respectively, of the National Research Council.

www.national-academies.org

---

*Resigned from the panel April 2, 2007.

Acknowledgments

THE PANEL to Review the Programs of the Bureau of Justice Statistics of the Committee on National Statistics (CNSTAT) is pleased to submit this interim report on options for conducting the National Crime Victimization Survey (NCVS). The work that leads to such a report always represents a collectivity—a devoted and talented staff of the National Research Council (NRC) and a set of volunteers, both panel members and those who met with the panel. Finally, the agency seeking advice from CNSTAT is key to the success of the endeavor.

The staff of the Bureau of Justice Statistics (BJS) has been exceptionally receptive to our external review of the agency's programs. Directed by Jeffrey Sedgwick, BJS has been generous in providing information and materials for the panel's consideration. Deputy director Maureen Henneberg has provided considerable assistance as the lead liaison between BJS and the panel, and fellow deputy director Allen Beck gave greatly of his time and expertise in interacting with the panel. Patrick Campbell, special assistant to the director, also participated in the public sessions of the panel's meetings. Michael Rand, chief of victimization statistics, deserves particular credit for leading a thorough and extremely useful review of the NCVS at the panel's first meeting. More than just cooperation is notable; in its meetings with the BJS staff, the panel observed clear devotion among the BJS staff to the quality and efficiency of the agency's statistical activities and a common purpose of serving the country well through its activities.

We greatly appreciate the work of the Justice Research and Statistics Association; its invitation to panel staff to attend the association's annual meeting in Denver in October 2006 was helpful in structuring the panel's work. Joan Weiss, executive director of the association, provided helpful comments and suggestions as the panel began its work.

In addition to the BJS staff, we gratefully acknowledge the other expert speakers who contributed to our plenary meetings: Kim English, research director, Division of Criminal Justice, Colorado Department of Public Safety; Mark Epley, senior counsel to the deputy attorney general, U.S. Department of Justice; Pat Flanagan, assistant division chief, Demographic Statistical Methods Division, U.S. Census Bureau; David Hagy, deputy assistant attorney general for policy coordination, Office of Justice Programs, U.S. Department of Justice; Douglas Hoffman, director, Center for Research, Evaluation, and Statistical Analysis, Pennsylvania Commission on Crime and Delinquency; Howard Hogan, associate director for demographic programs, U.S. Census Bureau; Krista Jansson, Home Office, United Kingdom; Ruth Ann Killion, chief, Demographic Statistical Methods Division, U.S. Census Bureau; Cheryl Landman, chief, Demographic Surveys Division, U.S. Census Bureau; Marilyn Monahan, chief of NCVS Branch, Demographic Surveys Division, U.S. Census Bureau; Jon Simmons, head of research analysis and statistics, Crime Reduction and Community Safety Group, Home Office, United Kingdom; and Philip Stevenson, statistical analysis center director, Arizona Criminal Justice Commission.

We extend our appreciation to directors of state statistical analysis centers for their participation in an informal survey about the use of victimization data in the states. This survey was led by panel member William Clements, who went far beyond the norm in his volunteering for such labor.

The study director of the panel was Daniel Cork, whose ability to absorb reams of technical, administrative, and organizational information about BJS and the Department of Justice earned the admiration of all panel members. His wisdom in assembling and integrating the writing of panel members and in structuring and writing the report was notable. The panel's work is conducted in cooperation with the NRC's Committee on Law and Justice (CLAJ). As senior program officer to this panel, CLAJ director Carol Petrie helped the panel integrate its work with prior studies and activities of the NRC concerning the Department of Justice. Agnes Gaskin, the senior program assistant, made sure meetings were organized and conducted in the professional manner that CNSTAT always achieves.

The Survey Research Center of the Institute for Social Research at the University of Michigan hosted a deliberative session of the panel at Ann Arbor in June 2007. Deborah Serafin, Rose Myers, and Kelly Smid handled the arrangements for the panel with grace and efficiency.

This report has been reviewed in draft form by individuals chosen for their diverse perspectives and technical expertise, in accordance with procedures approved by the Report Review Committee of the National Research Council (NRC). The purpose of this independent review is to provide candid and critical comments that will assist the institution in making the published report as sound as possible and to ensure that the report meets institutional standards for objectivity, evidence, and responsiveness to the study charge.

The review comments and draft manuscript remain confidential to protect the integrity of the deliberative process.

We thank the following individuals for their participation in the review of this report: Betsy Martin, retired, U.S. Bureau of the Census; Pat Mayhew, Crime and Justice Research Centre, Victoria University of Wellington, New Zealand; David McDowall, School of Criminal Justice, University at Albany, State University of New York; Henry N. Pontell, Department of Criminology, Law and Society and School of Social Ecology, University of California, Irvine; Callie Marie Rennison, Department of Criminology and Criminal Justice, University of Missouri–St. Louis; James F. Short, Sociology Department, Washington State University; and Alan M. Zaslavsky, Department of Health Care Policy, Harvard Medical School, Boston, Massachusetts.

Although the reviewers listed above provided many constructive comments and suggestions, they were not asked to endorse the conclusions or recommendations nor did they see the final draft of the report before its release. The review of the report was overseen by Philip J. Cook, Terry Sanford Institute of Public Policy, Duke University. Appointed by the NRC, he was responsible for making certain that an independent examination of the report was carried out in accordance with institutional procedures and that all review comments were carefully considered. Responsibility for the final content of this report rests entirely with the authoring panel and the institution.

Robert M. Groves, *Chair*
Panel to Review the Programs of the Bureau of Justice Statistics

Contents

List of Figures

List of Tables

List of Boxes

Executive Summary

THE BUREAU OF JUSTICE STATISTICS (BJS) of the U.S. Department of Justice, Office of Justice Programs (OJP), requested that the Committee on National Statistics (in cooperation with the Committee on Law and Justice) convene this Panel to Review the Programs of the Bureau of Justice Statistics. The panel has a broad charge to:

> examine the full range of programs of the Bureau of Justice Statistics (BJS) in order to assess and make recommendations for BJS' priorities for data collection. The review will examine the ways in which BJS statistics are used by Congress, executive agencies, the courts, state and local agencies, and researchers in order to determine the impact of BJS programs and the means to enhance that impact. The review will assess the organization of BJS and its relationships with other data gathering entities in the Department of Justice, as well as with state and local governments, to determine ways to improve the relevance, quality, and cost-effectiveness of justice statistics. The review will consider priority uses for additional funding that may be obtained through budget initiatives or reallocation of resources within the agency. *A focus of the panel's work will be to consider alternative options for conducting the National Crime Victimization Survey, which is the largest BJS program.* The goal of the panel's work will be to assist BJS to refine its priorities and goals, as embodied in its strategic plan, both in the short and longer terms. The panel's recommendations will address ways to improve the impact and cost-effectiveness of the agency's statistics on crime and the criminal justice system. [emphasis added]

BJS specifically requested that the panel begin its work by providing guidance on options for conducting the National Crime Victimization Survey (NCVS), one of many data series sponsored by BJS and one that consumes a large share (as much as 60 percent) of the agency's annual appropriations. This interim report responds to this request.

Since the survey began full-scale data collection in the early 1970s, the NCVS has become a major social indicator for the United States. Serving as a complement to the official measure of crimes reported to the police (the Uniform Crime Reporting [UCR] program administered by the Federal Bureau of Investigation), the NCVS has been the basis for better understanding the cost and context of criminal victimization. However, and particularly over the course of the last decade, the effectiveness of the NCVS has been undermined by the demands of conducting an increasingly expensive survey in an effectively flat-line budgetary environment. In order to keep the survey going in light of tight resources, BJS has reduced the survey's sample size over time, and other design features have been altered. When the survey began in 1972, the sample of addresses for interviewing numbered 72,000; in 2005, the NCVS was administered in about 38,600 households, yielding interviews with 67,000 people. Although this sample size still qualifies the NCVS as a large data collection program, occurrences of victimization are essentially a rare event relative to the whole population: many respondents to the survey do not have incidents to report when they are contacted by the survey. At present, the sample size is such that only a year-to-year change of 8 percent or more in the NCVS measure of violent crime can be deemed statistically to be significantly different from no change at all. In its reports on the survey, BJS has to combine multiple years of data in order to comment on change over time, which is less desirable than an annual measure of year-to-year change.

In approaching this work, the panel recognizes the fiscal constraints on the NCVS, but we do not intend to be either strictly limited by them or completely indifferent to them. Rather, our approach is to revisit the basic goals and objectives of the survey, to see how the current NCVS program meets those goals, and to suggest a range of alternatives and possibilities to match design features to desired sets of goals.

PRESERVING THE VICTIMIZATION MEASURE

There are no nationally available data on crime and victimization— collected at the incident level, with extensive detail on victims and the social context of the event—except those collected by the NCVS. It is this basic fact that is the strongest argument for the continuation and maintenance of the survey. Certainly, one option for the future of the NCVS—and the ultimate cost-reducing option—is to suspend or terminate the survey. It is an option that would have to be considered, if budget constraints require further reductions in sample size. To be clear, though, abandonment of the NCVS is not an option that we favor in any way.

Annual national-level estimates from the NCVS are routinely used in conjunction with the UCR to describe the volume and nature of crime in the United States. There is great value in having two complementary but nonidentical systems—the NCVS and the UCR—addressing the same phenomenon, for the basic reason that crime and victimization are topics that are too broad to be captured neatly by one measure. The police are not a disinterested party when it comes to characterizing the crime problem, and it is unwise to have data generated by the police as a sole measure of crime nationally. The UCR tells us little about the victims of crime; although its National Incident-Based Reporting System (NIBRS) has the potential to capture some of the detail currently measured by the NCVS, NIBRS has substantial limitations and remains incapable of providing national-level estimates after 20 years of implementation. Moreover, it is clear that a substantial proportion of crime is not reported fully and completely to law enforcement authorities. Thus, there remains a vital role for a survey-based measure that sheds light on unreported crime.

> **Recommendation 3.1:** BJS must ensure that the nation has quality annual estimates of levels and changes in criminal victimization.

The current design of the NCVS has benefited from years of experience, methodological research, and evaluation; it is a good and useful model that has been adopted by international victimization surveys as well as subnational surveys within the United States. The principal fault of the current NCVS is not a design flaw or methodological deficiency, or even that the design inherently costs too much to sustain, but rather—simply—that it costs more than is tenable under current budgetary priorities. In its present size and configuration, the NCVS can permit insights into the dynamics of victimization. However, in our assessment, the current NCVS falls short of the vibrant measure of annual change in crime that was envisioned at the survey's outset.

> **Finding 3.1:** As currently configured and funded, the NCVS is not achieving and cannot achieve BJS's legislatively mandated goal to "collect and analyze data that will serve as a continuous and comparable national social indication of the prevalence, incidence, rates, extent, distribution, and attributes of crime . . ." (42 U.S.C. 3732(c)(3)).

By several measures—comparison with the expenditures of foreign countries for similar measurement efforts or with the cost of crime in the United

States—the NCVS is underfunded. Accordingly, the panel recommends that BJS be afforded the budgetary resources necessary to generate accurate measures of victimization, which are as important to understanding crime in the United States as the UCR measure of crimes reported to the police.

> **Recommendation 3.2:** Congress and the administration should ensure that BJS has a budget that is adequate to field a survey that satisfies the goal in Recommendation 3.1.

OVERALL GOAL AND DESIGN CONSIDERATIONS

In considering historical goal statements of the NCVS, as well as new ones, we find three basic goals to be particularly prevalent and important, in addition to the previously expressed goal of maintaining annual national-level estimates of victimization that are independent of official reports to the police:

- Flexibility, in terms of both content (capability to provide detail on the *context and etiology of victimization* and to assess *emerging crime problems*, such as identity theft, stalking, or violence against and involving immigrants) and analysis (providing *informative metrics beyond basic crime rates*);

- Utility for gathering information on crimes that are not well reported to police or on *hard-to-measure constructs* (e.g., crimes against adolescents, family violence, and rape); and

- Small-domain estimation, including providing *information on states or localities*, which we think will be crucial to maximizing the utility of the NCVS and to building and maintaining constituencies for the survey.

In this report, we describe various design possibilities and their implications relative to these goals; however, we do not suggest one single path as the ideal for a redesigned NCVS. In part, this is because it is difficult to justify the case that our preferred set of NCVS goals is correct to the exclusion of all others; in part, it is because of the short time frame and the sequencing of this report (since it is inherently difficult to try to consider NCVS in isolation from the balance of BJS programs). But in large part we refrain from expressing a single, unequivocal path because the potential effectiveness and cost implications of some major design choices are simply unknown at this time.

We do think that it is critical to emphasize that even small changes to the design of a survey can have significant impacts on resulting estimates and the errors associated with them. Design changes made in the name of fiscal expedience, without grounding in testing and evaluation, are highly inadvisable. They risk unexplained changes in the time series and confusion among users.

>*Recommendation 4.1:* BJS should carefully study changes in the NCVS survey design before implementing them.

One potential cost-saving design choice is to change from asking respondents to recall and describe crime incidents in a 6-month window to using a 12-month window. This would entail contacting households once a year rather than twice (and, presumably, only 3 or 4 times if one chose to keep with the current regime of keeping households in the sample for 3.5 years). This would reduce the per-unit interviewing cost and free up resources to add additional sample addresses within each single year; 12 months is also the common reference period in victimization surveys in other countries. However, it could also increase problems of recall error by making respondents search their memories over a longer period. On its conceptual strengths and its use in comparable crime surveys in other western nations, we prefer a switch to a 12-month reference period as a cost-saving mechanism over options that would simply reduce the total sample size. That said, the empirical case for implementing this change is not completely clear and warrants up-to-date research. We note that such a move requires an overlap of designs over time to safely incorporate the change to 12 months.

>*Recommendation 4.2:* **Changing from a 6-month reference period to a 12-month reference period has the potential for improving the precision per-unit cost in the NCVS framework, but the extent of loss of measurement quality is not clear from existing research based on the post-1992-redesign NCVS instrument. BJS should sponsor additional research—involving both experimentation as well as analysis of the timing of events in extant data—to inform this trade-off.**

It is also the case that cost savings might be achieved by refining the NCVS sample stratification schemes. The current multistage cluster design of the NCVS automatically includes households sampled from counties and other geographic regions with large population sizes, clustering the remaining geographic areas by social and demographic information to produce similar strata from which the remaining sample is drawn. The composition of the sample is relatively slow to change with each decennial census, although effort is made to include some new housing stock by sampling from housing permit data. If the NCVS continues to be conducted by the Census Bureau (see "Collecting the Data," below), particular insight for altering the basic sample design and modifying sample strata based on an up-to-date sampling frame could come from interaction with the new American Community Survey (ACS). But, again, quantitative methodological research that could suggest exactly what benefits might or might not accrue is lacking.

> *Recommendation 4.7:* BJS should investigate changing the sample design to increase efficiency, thus allowing more precision for a given cost. Changes to investigate include:
>
> (i) changing the number or nature of the first-stage sampling units;
>
> (ii) changing the stratification of the primary sampling units;
>
> (iii) changing the stratification of housing units;
>
> (iv) selecting housing units with unequal probabilities, so that probabilities are higher where victimization rates are higher; and
>
> (v) alternative person-level sampling schemes (sampling or subsampling persons within housing units).

As early as 1980, the NCVS began the use of multiple response modes. Face-to-face personal interviews after the first contact with a sample household were replaced with interviews conducted by telephone, and—after the 1992 implementation of the full NCVS redesign—some interviewing began to be done by Census Bureau computer-assisted telephone interviewing (CATI) centers using a fully automated survey instrument. The NCVS path to automation has been somewhat complicated: full conversion to nonpaper survey questionnaires was achieved only in 2006, and—as part of the most recent round of cost reductions—BJS and the Census Bureau abandoned the use of the centralized CATI centers for NCVS interviews because anticipated cost savings never occurred. However, as redesign possibilities are considered, it is important that BJS continue to seek automation possibilities and not be limited to the NCVS traditional interview formats. A particular area of focus should be self-response options, such as computer-assisted self-interviewing (effectively, turning the interviewer's laptop around so that the respondent answers questions directly) or Internet response for interviews after several visits. As with the central CATI centers, cost savings from new modes of data collection are not guaranteed, but they may put the survey in good stead for implementing new topical modules and promoting high respondent cooperation. They can also serve to reduce overall respondent burden.

> *Recommendation 4.8:* BJS should investigate the introduction of mixed mode data collection designs (including self-administered modes) into the NCVS.

The NCVS is subject to the same pressures facing all household surveys in modern times, whether federal or private. It is increasingly difficult (and expensive) to obtain survey responses from persons or households in an age of cell phones, call waiting, and Internet chat. A significant fraction of survey costs are incurred to contact the most hard-to-find respondents. In considering design possibilities, it is important that BJS try to develop schemes that are relatively robust to declines in response rate, as such declines are virtually certain.

> *Recommendation 4.9:* **The falling response rates of NCVS are likely to continue, with attendant increasing field costs to avoid their decline. BJS should sponsor nonresponse bias studies, following current OMB guidelines, to guide trade-off decisions among costs, response rates, and nonresponse error.**

BUILDING AND REINFORCING CONSTITUENCIES

A continuing challenge for the NCVS is the development of constituencies with a strong interest in the data and their quality. The public is aware of the NCVS mainly due to one regular constituency—the media—and the spate of crime uptick or downtick stories that accompanies each year's release of NCVS and UCR estimates. Likewise, findings from topical supplements (such as racial dimensions of traffic stops, measured by the Police-Public Contact Survey supplement) typically get prominent press coverage. Official statistics, like other societal infrastructures, are often highly valued but rarely passionately promoted by day-to-day users. However, the long-term viability of the survey depends crucially on building and shoring up constituencies for NCVS products and on cultivating the survey's user base among researchers.

Small-Domain Estimates

The world has changed since the mid-1970s—computers are more powerful, data users are more sophisticated, and the demand for small-area geographic data is more insatiable. It is too strong to say that the NCVS can remain relevant *only* if it provides estimates for areas or populations smaller than the nation as a whole: state and local governments, which are among the most prodigious of NCVS users, continue to find national benchmarks very valuable. However, the survey will increasingly grow out of step with potential constituencies if it cannot be used to provide estimates for smaller areas.

Recommendation 4.5: BJS should investigate the use of modeling NCVS data to construct and disseminate subnational estimates of major crime and victimization rates.

This recommendation runs counter to the principal effect of one of our predecessor National Research Council (1976b) panel's recommendations— that the separate "impact city" victimization surveys that were originally part of the National Crime Surveys suite should be terminated. However, it is very much consistent with that previous recommendation's focus on an integrated set of estimates, including subnational geographies. These subnational estimates need not be exhaustive: expanding the sample to support estimates for the largest metropolitan statistical areas is a more sensible and cost-effective approach than a system for generating estimates for all 50 states. But they should permit insight on victimization for some smaller units than the nation as a whole. Small-domain estimates also refer to estimates by other social or demographic constructs, such as urbanicity (urban, suburban, or rural), in addition to the basic disaggregation by major race-ethnicity groups that is currently done.

With particular regard to the generation of small-domain estimates, it should be noted that enhancing the NCVS to better serve constituencies is not strictly a process of addition, in terms of sample size or implementation of a full supplemental questionnaire. In some important respects, user constituencies may best be served by more creative use of the current NCVS design. In the years since National Research Council (1976b) advocated eliminating the city surveys, statistical developments in small-domain estimation techniques have been considerable; hence, some small-domain estimates may be possible through modest investment by BJS in technical infrastructure for statistical modeling tasks.

In addition to small-domain modeling using NCVS data, it may also be useful to explore ways to strengthen victimization surveys conducted by states and localities. Currently, BJS operates a program under which it develops victimization survey software and provides it to interested local agencies; however, those agencies must supply all the resources (funds and manpower) to conduct a survey. An approach to strengthen this program would be to make use of BJS's organizational position within the U.S. Department of Justice. The bureau is housed in the Office of Justice Programs, the core mission of which is to provide assistance to state and local law enforcement agencies; it does so through the technical research of the National Institute of Justice and the grant programs of the Bureau of Justice Assistance (BJA), among others. We suggest that OJP consider ways of dedicating funds—like BJA grants, but separate from BJS appropriations—for helping states and localities bolster their crime information infrastructures through the establishment and regular conduct of state or regional victimization surveys. Such surveys would most likely involve cooperative arrangements with research organizations or local universities and make use of the existing BJS statistical analysis center infrastructure. This approach is analogous to the Behavioral Risk Factor Surveillance System (BRFSS) of the Centers for Disease Control and Prevention, and it is similar in its partnership arrangements to the Federal-State Cooperative Program for Population Estimates (FSCPE) of the Census Bureau.

> *Recommendation 4.6:* **BJS should develop, promote, and coordinate subnational victimization surveys through formula grants funded from state-local assistance resources.**

We discuss an extreme interpretation of this approach—wherein the "national" victimization survey would be effectively be the combination of the subnational surveys—in Chapter 4. However, we emphasize that we suggest that this BRFSS/FSCPE approach should be considered independent of (and as a complement to) the chosen design of the NCVS.

Topic Constituencies

The NCVS first added a topic supplement to the survey questionnaire in 1977, querying respondents on their perceptions of the severity of crime. Particularly since 1989, supplements have been an irregular part of the NCVS structure; the School Crime Supplement on school safety has been repeated six times and the Police-Public Contact Survey three times, with other supplements being (to date) one-time efforts.

A strong program of topic supplements is an important part of the NCVS, both because of the breadth of topics that may be handled and because the ability to quickly field questions on new topics of interest is a key advantage of survey-based collection compared with official records.

> *Recommendation 4.3:* **BJS should make supplements a regular feature of the NCVS. Procedures should be developed for soliciting ideas for supplements from outside BJS and for evaluating these supplements for inclusion in the survey.**

What is necessary regarding NCVS supplements is a more structured plan for their implementation, better exploration (and marketing) of sponsorship opportunities by other state and federal agencies, and greater transparency in real costs of conducting a supplement. Regardless of the overall design of the NCVS, the British Crime Survey offers an attractive model: a streamlined core set of questions combined with a planned, regular slot for topical content.

Recommendation 4.4: BJS should maintain the core set of screening questions in the NCVS but should consider streamlining the incident form (either by eliminating items or by changing their periodicity).

This would reduce respondent burden and allow additional flexibility for adding items to broaden and deepen information about prevalent crimes.

ATTENTION TO DATA QUALITY AND ACCESS

We make a series of recommendations that are agency-level in focus, aimed at better equipping BJS to understand its own products and to interact with its users. They are presented here in initial form because they are pertinent to the NCVS. We expect to expand on them in our final report on the full suite of BJS programs and products.

First, BJS currently receives periodic advice from the Committee on Law and Justice Statistics of the American Statistical Association (ASA). Although this input is certainly valuable, we think that BJS—and the NCVS in particular—would benefit from the commissioning of an ongoing scientific technical advisory board, such as is in place for other statistical agencies. This board should include subject matter, survey methodological, and statistical expertise; spots on the board are also a vehicle for strengthening stakeholder constituencies for the NCVS.

Recommendation 5.1: BJS should establish a scientific advisory board for the agency's programs; a particular focus should be on maintaining and enhancing the utility of the NCVS.

Several of our recommendations listed earlier identify gaps in existing research that must be filled to accurately inform trade-offs in design choices. More generally, the NCVS developmental work in the 1970s and the research conducted as part of the 1980s redesign effort are extensive, but we think that there is a paucity of recent methodological research making use of the post-1992-redesign NCVS instrument and techniques. BJS has already made some strides in fostering methodological research with its fellowship program, operated in conjunction with the ASA. We urge BJS to continue this work and to explore other creative ways to foster internal and extramural research using the NCVS and other BJS data sets, including graduate fellowships, as part of continuous efforts to assess the quality of NCVS estimates.

Recommendation 5.3: BJS should undertake research to continuously evaluate and improve the quality of NCVS estimates.

Conceptually, the survey-based NCVS is ideally suited (as the official record-based UCR is not) to study the dynamics of crimes that are emotionally or psychologically sensitive, such as violence against women, violence against adolescents, and stalking or harassment. We urge BJS to develop lines of research to ensure that such crimes are accurately measured on the NCVS instrument; these might include the testing of self-response options, such as audio computer-assisted interviewing.

> *Recommendation 3.3:* BJS should continue to use the NCVS to assess crimes that are difficult to measure and poorly reported to police. Special studies should be conducted periodically in the context of the NCVS program to provide more accurate measurement of such events.

The quality of NCVS data and its scientific rigor in measuring crime should always be the survey's primary goal and acknowledged as its principal benefit. However, for the purpose of cultivating constituencies and users for the survey, attention to the accessibility and the ease of use of NCVS data is also vitally important. Part of this work involves reevaluation of basic products and reports from the NCVS and expansion of the range of analyses based on the data, and it involves both in-house research by BJS and effective ties with other users and researchers.

> *Recommendation 5.2:* BJS should perform additional and advanced analysis of NCVS data. To do so, BJS should expand its capacity in the number and training of personnel and the ability to let contracts.

A necessary consequence of this recommendation is that the agency must expand its capacity, both in the number and training of personnel and the agency's ability to let contracts for external research.

> *Recommendation 5.4:* BJS should continue to improve the availability of NCVS data and estimates in ways that facilitate user access.

> *Recommendation 5.5:* The Census Bureau and BJS should ensure that geographically identified NCVS data are available to qualified researchers through the Census Bureau's research data centers, in a manner that ensures proper privacy protection.

In the case of this last recommendation, we understand that arrangements to place detailed NCVS data at the research data centers are under development; we state it here as encouragement to finalize the work.

COLLECTING THE DATA

It is important to note that some of the resource constraints on the NCVS are common to those on other important federal surveys, which have faced difficulties carrying out basic maintenance tasks like updating samples to reflect new census and address list information. The country needs a mechanism to alert itself to budget cuts that undermine the basic purposes of key federal statistical products.

> *Recommendation 5.6:* **The Statistical Policy Office of the U.S. Office of Management and Budget is uniquely positioned to identify instances in which statistical agencies have been unable to perform basic sample or survey maintenance functions. For example, BJS was unable to update the NCVS household sample to reflect population and household shifts identified in the 2000 census until 2007. The Statistical Policy Office should note such breakdowns in basic survey maintenance functions in its annual report** *Statistical Programs of the United States Government.*

Any review of a major survey program—particularly one carried out with an eye toward cost reduction—must inevitably raise the question of the agent that collects the data: could survey operations be made better, faster, or cheaper by getting some other organization to carry out the survey? In this case, the U.S. Census Bureau's involvement with the NCVS predates the formal establishment of the survey, as the Census Bureau convened planning discussions and conducted NCVS pilot work.

The optimal decision on who should do the data collection for the NCVS will depend on the weight that one puts on desired objectives for the survey. For instance, an extremely strong weight on flexibility and quick response to emerging trends might argue against the Census Bureau, where implementation of a supplement can be made time-consuming through detailed cognitive testing (which ultimately improves the quality of the questions but can be slow) and passage through bureaucratic channels (e.g., clearance by the Office of Management and Budget, as required of all federal surveys). However, dominant weight on maintaining high response rates and drawing from the experience of other large, ongoing surveys would suggest that staying with the Census Bureau is the best course. Just as we do not offer a single design path for the NCVS, we do not find justification for offering a conclusion on "Census Bureau" or "not Census Bureau." Based on the advantages and disadvantages, we suggest that "privatizing" the NCVS is not the panacea for high survey costs that some may believe it is. We have been provided no way of estimating the various costs associated with switching NCVS data collection agents; however, it is altogether appropriate to consider means of getting detailed and specific answers to these questions.

In the interim, we suggest that the Census Bureau would benefit both BJS and itself itself by providing greater transparency in true survey costs.

Recommendation 5.7: **Because BJS is currently receiving inadequate information about the costs of the NCVS, the Census Bureau should establish a data-based, data-driven survey cost and information system.**

We further suggest that BJS consider a design competition—providing some funds for bidders to specify in detail how they would conduct a victimization survey. This design competition would effectively compensate bidders for their time in developing proposal specifications, but it should be run with a statement that a formal request for proposals *may* result from the competition (and not that it will definitely occur).

Recommendation 5.8: **BJS should consider a survey design competition in order to get a more accurate reading of the feasibility of alternative NCVS redesigns. The design competition should be administered with the assistance of external experts, and the competition should include private organizations under contract and the Census Bureau under an interagency agreement.**

– 1 –

Introduction

THIRTY-FOUR YEARS AGO, a panel convened by the National Research Council (NRC) began work as requested by the U.S. Department of Justice, reviewing the department's survey-based system to measure criminal victimization. The panel's work was not preparatory to data collection, helping to design the survey from scratch, nor was it a review of a long-established program, as the full survey had been in the field for only two years. Instead, the panel's work was to serve as an early course correction—a review of the objectives of what was then a suite of related victimization surveys, coupled with short- and long-term guidance on retooling the surveys to meet specified objectives.

Several of the recommendations in the previous panel's final report *Surveying Crime* (National Research Council, 1976b) were adopted fairly rapidly, including the elimination of victimization surveys of commercial establishments and of selected large cities.[1] The National Crime Survey was established as the name of the survey of sampled households. The report also contributed to a discussion of long-term redesign options starting in the early 1980s—initiated by the Department of Justice and executed by a consortium of academic and government survey researchers—that culminated in a reengineered (and renamed) National Crime Victimization Survey (NCVS) in 1992.[2]

---

[1]The 1976 panel's principal recommendations are listed in Appendix B.

[2]The nomenclature of the survey—and that of its sponsoring agency—has shifted over time. The most preliminary study plan for the survey in 1970 (Work, 1975) described the new program as the *National Victimization Survey*. When data collection began, the program was described as the *National Crime Surveys*—plural, since the program included national and city-

As presently implemented, the NCVS is a major national household survey using a rotating panel sample of addresses: after an address is chosen for the survey, each person age 12 or older in the household at that address is interviewed seven times at 6-month intervals. The first interview with a household is always done by personal visit, but subsequent interviews may be done by telephone if a number is available. The first portion of the post-1992 NCVS is a screening questionnaire, using detailed questions to elicit counts and basic information about crime victimization incidents in the preceding 6 months. An incident report is then prepared for each incident detected in the screener, including a battery of questions on the context of each event. (The operations of the NCVS are described in more detail in Appendix C.) In 2005, the NCVS was administered to approximately 38,600 households, yielding interviews with 67,000 people. Sponsored by the Bureau of Justice Statistics (BJS) of the U.S. Department of Justice, the field collection of the NCVS is performed by the U.S. Census Bureau.[3] Annually, BJS publishes reports and summary tables from the NCVS in two continuing report series, *Criminal Victimization* and *Crime and the Nation's Households* (see, e.g., Catalano, 2006; Klaus, 2007). BJS also uses NCVS data as the basis for periodic or one-shot reports on a wide array of topics and victimization types, including carjacking (Klaus, 1999, 2004), firearm use in crime (Rand, 1994; Zawitz, 1995), perceptions of neighborhood crime (DeFrances and Smith, 1998), victimization of college students (Baum and Klaus, 2005), and workplace violence (Warchol, 1998).

Over its 35-year history—major highlights of which are listed in Box 1-1—the NCVS has been a uniquely valuable source of information on crime. Intended to shed light on the "dark figure of crime"—the phrase coined by Biderman and Reiss (1967) to describe criminal incidents that are not reported to police—it is frequently used in conjunction with data from the Uniform Crime Reporting (UCR) program, through which the Fed-

level surveys of businesses as well as national and city-level household surveys. The national household survey component—originally referred to as the *National Crime Panel* program—was the only part to survive into the late 1970s and became known as the *National Crime Survey* (singular), commonly abbreviated as NCS. This remained the standard nomenclature until the 1992 switch to **National Crime Victimization Survey** or NCVS. In this report, we use NCS in direct quotations when applicable, but generally use NCVS as the descriptor.

[3]Another terminological note is in order for references to BJS. The organizational unit responsible for developing and analyzing the survey was—as of 1970—known as the *National Criminal Justice Statistics Center*. By mid-decade, it was dubbed the *National Criminal Justice Information and Statistics Service* (NCJISS) and was a constituent component of the *Law Enforcement Assistance Administration* (LEAA) of the U.S. Department of Justice. In 1984, reauthorization legislation dispersed the functions of the LEAA throughout a new **Office of Justice Programs** (OJP) in the Justice Department, overseen by an assistant attorney general. The statistical and data-gathering functions of the former NCJISS were vested in the new **Bureau of Justice Statistics** (BJS). As with references to the NCVS, we tend to use "BJS" to describe the agency, regardless of the exact date in question, except in direct quotations.

Box 1-1   Major Steps in the Development of the National Crime
Victimization Survey

- **1967:** The President's Commission on Law Enforcement and Administration of Justice
  (1967) issues its final report. The report contrasts results of its 10,000-household
  National Survey of Criminal Victims (conducted by the National Opinion Research
  Center), as well as city-level victim surveys for Washington, Chicago, and Boston
  (conducted by the Bureau of Social Science Research and the University of Michigan
  Survey Research Center), with the FBI's Uniform Crime Reports. Concluding that lack
  of information on offenses not reported to police makes it difficult to develop useful
  crime policy, the commission recommends the development of a nationwide crime
  victimization survey. Procedurally, the commission's recommendations also lead to the
  1968 establishment of the Law Enforcement Assistance Administration in the U.S.
  Department of Justice.

- **1968:** Initial discussions between the LEAA and the Census Bureau take place on a
  national survey of crime victims. The Census Bureau conducts two staff workshops on
  the scope and basic design of such a survey, resulting in a conference report (U.S.
  Bureau of the Census, 1968).

- **1970–1971:** *Pretesting*—small-scale tests, most involving reverse record checks
  (interviews with known victims of crimes reported to police), are conducted in
  Washington, DC; Akron, Cleveland, and Dayton, Ohio; Baltimore, Maryland; and San
  Jose, California. Testing in San Jose and Dayton included the fielding of surveys to a
  sample of about 5,000 households in each site.

- **1971–1972:** *Initial implementation as supplement*—victimization questions are added
  to the Census Bureau's Quarterly Household Survey. The first such nationwide
  sample, in January 1971, had a sample size of 15,000 housing units and asked
  respondents to recall events within a 12-month reference period. (The same
  supplement was used in 1972, using a 6-month reference period.) Subsequent
  iterations were used to refine the design of questionnaires and instructions. These
  surveys-as-supplements were intended only as an experiment and not the basis for
  published reports.

- **July 1972:** The new National Crime Surveys—jointly developed by the LEAA and the
  Census Bureau—are fielded. The core National Crime Panel is a sample of 72,000
  households; one-sixth of the sample were to be interviewed monthly and then again
  at 6-month intervals. The suite of surveys also includes a national sample of 15,000
  businesses, as well as a sample of 12,000 households and 2,000 businesses in each of
  26 major cities, to support local-area estimates of victimization. These "Cities Surveys"
  begin in 8 "impact cities" and are fielded in sets of the cities over the next 3 years.

- **1976:** The Panel on the Evaluation of Crime Surveys of the National Research Council
  (1976b) releases its final report, *Surveying Crime*. The major recommendations of the
  panel are summarized in Appendix B; among the recommendations is that the Cities
  Surveys be abandoned in order to support the core national household sample.

- **1986:** *Redesign phase-in*—following the recommendations of a redesign consortium,
  small-scale changes to the survey and its instruments are implemented, with the larger
  scale changes originally intended to be implemented in 1989.

Box 1-1 (continued)

- **1988–1989:** *Redesign pretest*—a three-wave national pretest of the redesigned victimization survey is fielded, using a 6-month reference period and bounded data.

- **January 1989:** A 5 percent test group of respondents receive the redesigned survey questionnaire, including a revised set of screener questionnaires. The first major supplement to the victimization survey—the School Crime Supplement (SCS)—is also implemented and is conducted through June 1989.

- **1990:** Gradual phase-in of new questionnaire continues, with another 5 percent added to the sample.

- **July 1991:** *Renaming*—the survey name is changed to National Crime Victimization Survey (NCVS), to refer to the data collected using the new questionnaire.

- **January 1992:** *Conversion*—the sample receiving the redesign/NCVS questionnaire is elevated to 50 percent, providing a year of half-samples using the old and new forms. In July 1993, the redesigned NCVS questionnaire becomes the norm and the former NCS questionnaire is discontinued. For calendar year 1992, the Survey of Public Participation in the Arts (funded by the National Endowment for the Arts) is appended to the NCVS as a supplement.

- **Mid-1990s:** Context variables on public housing, college or university housing, and attendance or employment at colleges or universities are added to the screening questions. The SCS is replicated, with joint sponsorship by BJS and the National Center for Education Statistics, in 1995. Another supplement—the Police-Public Contact Survey (PPCS) is first fielded on a pilot basis (for outgoing rotation households) in May–July 1996.

- **May 1997:** BJS elects to suspend production of state-level tabulations (beginning with 1996 data) and preliminary estimates previously published in *Crime in the Nation's Households* (beginning with 1997 data).

- **1998–1999:** A parallel, 12-city survey conducted only by random-digit dialing is collected from February through May 1998; the survey is conducted by the Census Bureau and sponsored by BJS and the Office of Community Oriented Policing Services. The SCS and PPCS supplements are repeated in 1999, and hate crime questions are introduced in the main NCVS in the same year.

- **2001:** Questions on cybercrime are added to the NCVS, as are incident-report questions that relate to a Workplace Risk Supplement (requested by the National Institute for Occupational Safety and Health) fielded in 2002.

- **January 2003:** Race and ethnicity questions are revised, pursuant to the 1997 directive issued by the Office of Management and Budget.

- **2004:** Identify theft questions are added to the NCVS.

- **January 2006:** The Supplemental Victimization Survey (SVS), focusing on stalking, is fielded from January to June.

eral Bureau of Investigation compiles counts of crimes reported to law enforcement agencies nationwide (see Appendix D for additional detail on the UCR). However, it has also served a vital function by focusing attention on the characteristics of victims of crime, including the nature of those incidents and how people respond to them.

Highly influential in shaping contemporary knowledge and theories of criminal victimization, NCVS data have informed assessments of victimization risk, victim-offender relationships, weapon use, and injury in violent crime; the costs of violent and property victimization; and the degree to which crimes are reported to the police (and the reasons for reporting or not reporting). Through the basic survey and topical supplements, NCVS data have also offered insights on new types of criminal behavior such as identity theft (Baum, 2007) as well as violence types of increasing policy importance (e.g., intimate partner violence and crime in school settings). Significantly, the NCVS has also served "as a model for victimization surveys implemented throughout the world" (as well as victimization studies in individual states or cities) "because it incorporates many innovative methodological protocols that enhance its ability to produce reliable estimates of the nature and extent of criminal victimization" (Rennison and Rand, 2007:17). (Appendix E describes some of these foreign and state-level victimization surveys.)

Although the history of the NCVS has been marked by achievement, it has also been marked—almost from the survey's inception—by a certain level of unanticipated changes in design through budget cuts. The survey has been subject to a lengthy string of sample size reductions and changes in procedure, typically in the interest of reducing costs. Many of these are itemized in Box 1-2. In recent years, the problems of keeping the NCVS operational have been particularly acute because BJS—the survey's sponsor—has been effectively flat-funded since 2000. As shown in Figure 1-1, BJS—like many federal agencies—has consistently received less in total appropriations than requested, but appropriated funds have been particularly tight over most of the past decade. However—even with all of the cost-cutting measures described in Box 1-2—the cost of implementing the NCVS has continued to grow. In part, this is attributable to the fixed costs of converting the survey to electronic forms to facilitate computer-assisted interviews and to increased labor costs, but it is also due to the increasing difficulty of contacting survey respondents and gaining their cooperation.

For BJS, the net consequence of these fiscal constraints has been the painful trade-off decisions required to support a full suite of justice-related data collections while a single program has consumed the bulk of the agency's limited resources. Aside from victimization, BJS maintains major, core data collection efforts in the areas of corrections (including censuses and surveys of jails and prisons as well as data on probation and parole programs), law enforcement management and administration, courts and sen-

Box 1-2 Notable Reductions in the National Crime Victimization Survey

- **1977:** The commercial component of the National Crime Surveys—including measurement of commercial burglary and robbery—is eliminated, principally for budgetary reasons but also due to the inadequacy of the business sampling frame.
- **January 1980:** Personal visit interviewing is reduced and telephone interviewing introduced. Interview 1 with a household is still required to be conducted in person, but interviews 2, 4, and 6 are directed to be conducted by telephone to the extent possible.
- **June 1984:** A sample size cut is made, resulting in the reduction of nonself-representing strata from 220 to 153 and a 20 percent reduction in the larger of the 153 self-representing primary selection units. According to the Demographic Surveys Division, U.S. Census Bureau (2007b:25), this cut was imposed "so that funds could be redirected to pay for redesign research."
- **February 1986:** Due to budget problems, reinterview activities are suspended for one month and reduced by half for seven additional months.
- **March 1986:** To reduce costs, the protocol for personal visit interviewing rather than by telephone is revisited. Telephone interviewing for rotation groups 3 and 7 is increased, with only interviews 1 and 5 slated for personal visit.
- **Second half, 1992:** Due to funding constraints, discussions begin with the Census Bureau on possible ways to reduce the costs of the NCVS. To cut costs, reinterviews were stopped for one month in August 1992, and a 10 percent sample size reduction was implemented in October 1992 (affecting both the groups receiving the NCVS and NCS questionnaires).
- **October 1996:** A 12 percent sample cut is imposed, due to budget constraints. The cut came after a summer of some computer-assisted telephone interviewing (CATI) follow-up work (for noninterviews) being suspended due to a budget impasse and was suggested by the Census Bureau (along with two other options) as a "two-year plan."
- **July 2000:** Lifestyle and home protection questions are dropped from the screener portion of the questionnaire.
- **April 2002:** The NCVS sample is reduced by 4 percent.
- **January 2006:** The NCVS sample reduced by 16 percent.
- **July 2007:** Due to budget constraints, three major changes are slated to occur in summer 2007: (1) use data from the first interview—previously withheld as a "bounding" case—in annual estimates, (2) implement a 14 percent sample cut, as a balance for using the bounding first interviews, and (3) suspend all computer-assisted telephone interviewing from Census Bureau call centers (however, field interviewers may still use the telephone to conduct their scheduled interviews).

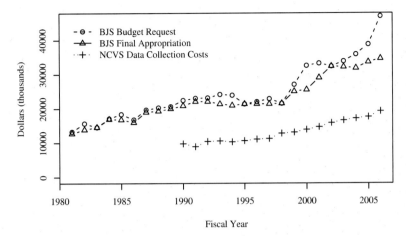

Figure 1-1 Bureau of Justice Statistics budget requests and final total appropriations, 1981–2006, and National Crime Victimization Survey data collection costs, 1990–2006

NOTE: Budget request and appropriations figures include both base and program costs. Final appropriations reflect amounts after any applicable budget recission or across-the-board cut. NCVS data collection costs exclude additional costs for developing and respecifying sample based on new decennial censuses, as well as costs associated with the automation (conversion to computer-based administration) of the survey.

SOURCE: Data provided by the Bureau of Justice Statistics as briefing for the panel's first meeting in February 2007.

tencing (including series on both civil and criminal cases at the state level, as well as data on all steps of processing in federal criminal cases), and expenditure and employment in the justice system. However, during fiscal years 2001 through 2006, the NCVS consumed at least 51.2 percent (in both 2001 and 2002) and as much as 64.0 percent (in 2004) of the total BJS appropriations. For the American public, the consequence has been the degradation of a key social indicator: rising data collection costs have led BJS to reduce the sample size of the survey and induce other less visible cost-cutting measures. Over the past decade, two trends—the diminishing NCVS sample size and generally low and decreasing estimated overall victimization rates—have combined, with the result that only large percentage changes in violent crime victimization rates—at least 8 percent—would be a statistically significant year-to-year change.[4] As noted in the U.S. Office of Management and

---

[4]This figure is from a presentation by Michael Rand, BJS, at the panel's first meeting; for 2005, the violent crime victimization rate was estimated as 21.2 per 1,000 population, and the 95 percent confidence interval of this rate is ±8.1 percent. Since 2000, the estimated annual

Budget (2007:8) annual review of statistical program funding, "cost cutting measures applied to the NCVS continue to have significant effects on the precision of the estimates—year-to-year change estimates are no longer feasible and have been replaced with two-year rolling averages" in BJS reports on victimization.

The predecessor panel offered an early course correction to the still-new National Crime Survey, noting that (National Research Council, 1976b:9–10):

> A survey—or any measurement process—should be designed to meet stated objectives in so far as feasible. Evaluation of the survey or the measuring instrument should in large part be an assessment of the degree to which it meets or can meet objectives.

That panel's work was furthered by the NCS redesign consortium in the 1980s, culminating in a phased-in implementation of a redesigned NCVS in 1992. The NCVS and its design has benefited from a steady stream of methodological research and refinement since its inception; now, some 35 years later and recognizing the challenges facing the NCVS and other survey data collections, the question of whether the survey is meeting its goals and user base needs is ripe for review.

1–A CHARGE TO THE PANEL

The Bureau of Justice Statistics requested that the National Academies' Committee on National Statistics (in cooperation with the Committee on Law and Justice) convene this Panel to Review the Programs of the Bureau of Justice Statistics with a broad charter to review the full suite of BJS programs. The panel has been tasked to review all of the agency's varied data collections, with an eye toward gaps in substantive coverage that should be filled as well as collections that should be suspended or dropped.

The full charge to the panel is to:

> examine the full range of programs of the Bureau of Justice Statistics (BJS) in order to assess and make recommendations for BJS' priorities for data collection. The review will examine the ways in which BJS statistics are used by Congress, executive agencies, the courts, state and local agencies, and researchers in order to determine the impact of BJS programs and the means to enhance that impact. The review will assess the organization of BJS and its relationships with other data gathering entities in the Department of Justice, as well as with state and local

---

violent crime victimization rates have dipped from 27.9 to 21.2 per 1,000, and the 95 percent intervals have been in excess of ±7 percent for each of those annual estimates. See Figure C-1 in Appendix C.

governments, to determine ways to improve the relevance, quality, and cost-effectiveness of justice statistics. The review will consider priority uses for additional funding that may be obtained through budget initiatives or reallocation of resources within the agency. *A focus of the panel's work will be to consider alternative options for conducting the National Crime Victimization Survey, which is the largest BJS program.* The goal of the panel's work will be to assist BJS to refine its priorities and goals, as embodied in its strategic plan, both in the short and longer terms. The panel's recommendations will address ways to improve the impact and cost-effectiveness of the agency's statistics on crime and the criminal justice system. [emphasis added]

Given the prominence of the NCVS in BJS operations—and its dominance of BJS budget resources—the panel was specifically asked to evaluate options for conducting the NCVS in our first year of work,[5] before turning to the agency's data collections related to other areas, like corrections and judicial processing. Consistent with this principal task of the report, it is important to make clear that this report is not intended to be a complete sourcebook on the NCVS; it is neither a full procedural history of the survey, a complete literature review of its uses, nor a detailed operational plan for any specific alternative design.[6] This report is also not meant to revisit in full detail the comprehensive redesign efforts that culminated in the fielding of the new NCVS instrument in 1992. That redesign effort is reported thoroughly by Biderman et al. (1986) and Bureau of Justice Statistics (1989) and major issues faced in the redesign are also summarized by Skogan (1990). It generated a large body of valuable methodological research, much of which exists in technical BJS and Census Bureau memoranda. We do not reprise that work in detail in this report, but the recommendations we make do suggest the need for an ongoing evaluation system to produce a similar, rich body of updated methodological work to inform future NCVS design changes.

Due to the nature of our panel's charge, it is also essential to underscore that this is a first-stage or interim report. As we discuss further in Section 3–D, the methodological focus of this report means that we do not attempt as exhaustive a listing of constituencies and uses for the NCVS as our overall charge suggests; we intend to consider a more complete assessment of the user base in our final report. Furthermore, this report does not and is not intended to provide comprehensive treatment of all BJS programs, nor

---

[5] In "Strengthening Federal Statistics," Appendix 4 of the president's proposed budget for fiscal year 2008, one of two initiatives specifically referenced in the BJS budget request is "a redesign of the National Crime Victimization Survey based on anticipated recommendations from the Committee on National Statistics of the National Research Council," i.e., this panel.

[6] The two volumes edited by Lehnen and Skogan (1981, 1984) are an important resource on the early history of the survey.

does it completely reconcile NCVS components with other BJS data collection activities. As this panel continues the work of suggesting data collection priorities based on a fuller review of the suite of BJS programs, it is possible that some NCVS-related issues will have to be revisited in the final report.

1–B   REPORT CONTENTS

Following this introduction, Chapter 2 reviews the current and historical goals of the NCVS, describing the various roles that the survey has been expected to fulfill. Chapter 3 then uses the goals of the NCVS to describe the major technical and operational issues surrounding the current conduct of the survey. In Chapter 4, we lay out several sets of goals for a reconfigured NCVS and describe the design choices that follow directly from particular sets of goals and priorities. We close in Chapter 5 with a look forward, making recommendations on how best to calibrate a victimization measurement system for the 21st century.

Appendix A reproduces the panel's recommendations for ease of reference and—as previously noted—Appendix B recounts the principal findings and recommendations of National Research Council (1976b). As a reference for the reader and to enhance the flow of the main text, we have placed detailed descriptions of the NCVS and related programs in appendixes. Appendix C describes the sample design of the NCVS, as well as the interviewing procedures for the survey and the content of the survey instrument. In Appendix D, we describe other data resources that are available for studying aspects of criminal victimization; a major portion of that appendix describes the Uniform Crime Reporting program of the Federal Bureau of Investigation which—as the national inventory of crimes reported to police—is the data source to which the NCVS is most commonly compared for the purpose of assessing crime trends in the United States. Finally, in Appendix E, we describe other existing victimization surveys, such as those conducted in foreign countries (particularly the British Crime Survey) and those that have periodically been fielded in American states and cities.

– 2 –

Goals of the National Crime Victimization Survey

I T IS EASY TO UNDERESTIMATE how little was known about crimes and victims before the findings of the National Crime Victimization Survey (NCVS) became common wisdom. In the late 1960s, knowledge of crimes and their victims came largely from reports filed by local police agencies as part of the Federal Bureau of Investigation's (FBI) Uniform Crime Reporting (UCR) system, as well as from studies of the files held by individual police departments. At the time, UCR coverage of the nation was far from complete and—in any event—almost all of the information it gathered came in highly aggregated form. Agencies sent in monthly totals for crimes in seven categories, which the FBI later published as yearly totals. The UCR program produced national data on the characteristics of specific incidents for only one crime type (homicide, through the Supplementary Homicide Reports). It produced no data on characteristics of victims, the costs or injurious consequences of crime, or the circumstances in which crimes occurred. Data on offenders came separately, in summary descriptions of the age, sex, and race of persons who were arrested, and national coverage of arrestees was even more incomplete than incident reporting. Furthermore, at that time—before the dawn of the information technology revolution—police recordkeeping was cumbersome and in many places haphazard, and it was widely feared that crime data were subject to manipulation if not misreporting, for political and organizational reasons.

Criminologists understood that there existed a "dark figure" of crime consisting of events not reported to the police (Biderman and Reiss, 1967). Scattered studies of victimization, principally in Europe, suggested that the dark figure might be both very large and highly selective, raising important questions about what could be learned from crime data.

Crime surveys promised to provide an alternative, often better, and certainly more adaptable source of data on crimes and victims, one that could shine new light on the dark figure. Surveys commissioned by the President's Commission on Law Enforcement and Administration of Justice (1967) during the 1960s led toward the NCVS, which was further shaped by an extensive program of methodological research during the 1970s. As it developed, the survey proved to be a source of several important streams of new data, giving analysts more flexibility in using them to address policy and research questions.

In evaluating and assessing any program, it is essential to consider the program's goals: what the program is intended to do and how objectives may have changed over time. Accordingly, this chapter reviews the fundamental goals of the NCVS, starting with a review of goal statements that have been advanced over the years (Section 2–A). In subsequent sections, we discuss some of the major themes in these goal statements in more detail: NCVS as a source of data on victims of crime (2–B), as counterpart to the UCR and a check on crimes reported to the police (2–C), as a tool for analytic flexibility in studying different crime types (2–D), as a source for information on special topics in crime (2–E), and as a response to specific legal mandates (2–F). We return to discussion of specific goals at the close of Chapter 3, after surveying major issues facing the NCVS.

2–A HISTORICAL GOAL STATEMENTS OF THE NCVS

As part of its comprehensive review, the President's Commission on Law Enforcement and Administration of Justice (1967:38) observed that "one of the most neglected subjects in the study of crimes is its victims: the persons, households, and businesses that bear the brunt of crime in the United States." The commission argued:

> Both the part the victim can play in the criminal act and the part he could have played in preventing it are often overlooked. If it could be determined with sufficient specificity that people or businesses with certain characteristics are more likely than others to be crime victims, and that crime is more likely to occur in some places than in others, efforts to control and prevent crime would be more productive. Then the public could be told where the risks of crime are greatest. Measures such as preventive police patrol and installation of burglar alarms and special locks could then be pursued more efficiently and effectively. Individ-

uals could then substitute objective estimation of risk for the general apprehension that today restricts—perhaps unnecessarily and at best haphazardly—their enjoyment of parks and their freedom of movement on the streets after dark.

To get an initial reading of such information, the commission contracted with the National Opinion Research Center to conduct a National Survey of Criminal Victims. "The first of its kind conducted on such a scope," the crime commission's survey reached a sample of 10,000 households. The survey results sufficed to demonstrate to the commission that "for the Nation as a whole there is far more crime than ever is reported" (President's Commission on Law Enforcement and Administration of Justice, 1967:v):

> Burglaries occur about three times more often than they are reported to police. Aggravated assaults and larcenies over $50 occur twice as often as they are reported. There are 50 percent more robberies than reported. In some areas, only one-tenth of the total number of certain kinds of crimes are reported to the police.

Based on these findings, the commission's final report recommended the creation of a National Criminal Statistics Center, a "major responsibility" of which "would be to examine the Commission's initial effort to develop a new yardstick to measure the extent of crime in our society as a supplement to the FBI's Uniform Crime Reports" (President's Commission on Law Enforcement and Administration of Justice, 1967:x).[1] That is, in addition to obtaining information from the oft-neglected perspective of victims of crime, the commission envisioned what would become the NCVS as a complementary measure to the crimes-reported-to-police measures of the UCR program. Notably, the commission emphasized this new data collection vehicle as a complement to, and not a replacement for, the UCR. Elsewhere in the report, the commission reiterated that "the development of public surveys of victims of crime . . . can become a useful supplementary yardstick" and emphasized (President's Commission on Law Enforcement and Administration of Justice, 1967:31) that:

> what is needed to answer questions about the volume and trend of crime satisfactorily are a number of different crime indicators showing trends over a period of time to supplement the improved reporting by police agencies.

Furthermore, the commission set forth a more ambitious goal for victimization surveys: providing information on the nature and levels of crime for

---

[1] A previous National Commission on Law Observance and Enforcement (the Wickersham Commission) made a similar call for criminal statistics program administered by an independent, central statistical agency; however, that recommendation languished (Cantor and Lynch, 2000:95).

areas smaller than the nation as a whole. "The Commission believes that the Government should be able to plot the levels of different kinds of crime in a city or a State as precisely as the Labor Department and the Census Bureau now plot the rate of unemployment. Just as unemployment information is essential to sound economic planning, so some day may criminal information help official planning in the system of criminal justice" (President's Commission on Law Enforcement and Administration of Justice, 1967:x).

As described in Box 1-1, the crime commission's report led to discussions between the U.S. Department of Justice and the Census Bureau on implementing a national survey of victims. An initial planning conference held by the Census Bureau articulated nine possible goal statements for the new data collection program (U.S. Bureau of the Census, 1968):

- Provide an independent calibration for the UCR;

- Provide a measure of victim risk;

- Enable a shift in concentration in the criminal justice system from the offender to the victim;

- Provide an indicator of the crime problem outside those indicators generated by police activity;

- Serve as an index of changes in reporting behavior in the population;

- Provide an indicator of the social "outlook" in the population as well as an indicator of society's definitions of crimes;

- Serve as a basis for the study of granting of compensation to victims;

- Serve as a statistic to determine the degree of involvement by the victim; and

- Serve as a measure of public confidence in police.

In 1970, the first proposed study plan for a national victimization survey refined this list and suggested that the survey's primary purpose would be "to measure the annual change in crime incidents for a limited set of major crimes and to characterize some of the socio-economic aspects of both the reported events and their victims" (reprinted in Work, 1975:220).

An unattributed document titled "Objectives of the National Crime Survey"—dated "circa 1976"[2] and apparently an internal Law Enforcement

---

[2]The document—provided by BJS to the panel as part of a briefing package for its first meeting—refers briefly to "the Academy report" and questions that "the Academy raised," referring to *Surveying Crime* (National Research Council, 1976b). However, it also includes reference to the commercial surveys that were part of the original National Crime Surveys but were dropped on the basis a recommendation in National Research Council (1976b). Hence, it appears to postdate the National Research Council report but not by a great deal. The language in this document is different from—and more speculative than—the formal statements of LEAA

Assistance Administration (LEAA) document—is broad in scope. It declares that "there is only one unchanging objective" of the survey but describes that objective in a manner that is different from past versions: "to learn about crime and criminal justice and closely related phenomena, that which can only (or best) be learned from a large representative National survey." This statement emphasizes the victimization survey's *methodological* independence from the UCR and official reports to police. Indeed, the document resists "the conceptual subjection of the NCS to the UCR": "the ability of the NCS to supplement, complement and enhance the UCR does exist," but the survey's inherent capability for "much broader and more varied approaches to crime and its consequences" is inherently limited "to the extent the NCS limits its conceptualization to the UCR framework." Accordingly, "it is not an objective of the NCS to measure simply what the UCR does not measure, nor to measure 'better' what the UCR does measure." The LEAA document notes three corollaries to this "unchanging objective":

1) to continuously pursue a program of conceptual and methodological research on a scale necessary to support such a large scale research effort and to maximize its capacity to inform and enlighten;

2) to fully exploit the capacity of the survey to inform and enlighten through a program of competent, creative, and thorough substantive analytical research;

3) to disseminate research findings to researchers, practitioners, policymakers and the general citizenry in forms most appropriate to insuring that the findings become part of the working knowledge of each audience.

The document also specified that certain statements should *not* be goals of the victimization survey. Although "good NCS data for metropolitan areas" might have "unique and valuable research potential," the document argues that "the NCS is not a suitable mechanism for informing local law enforcement planners and administrators about crime within their jurisdiction." Likewise, the survey "cannot be, and must not be promoted as being, a vehicle for evaluating local programs aimed at reducing crime or improving the criminal justice system."

Following the publication of *Surveying Crime* (National Research Council, 1976b), the Law Enforcement Assistance Administration (1981) listed seven formal "objectives of a publicly funded nationwide statistical series on victimization," emphasizing that the objectives are "current" and hence subject to revision:

---

objectives issued in 1975 (Work, 1975) and 1977 (Law Enforcement Assistance Administration, 1981), before and after the National Research Council report.

- To provide trend data that will serve as a set of continuous and comparable national social indicators for the rate of victimization for selected crimes of violence and crimes of theft and for other factors related to crime and victimization in support of national criminal justice policy and decisionmaking and in support of informed public discussion.

- To conduct a program of conceptual and methodological research that will improve the victimization surveys in response to the National Academy of Sciences evaluation, including refinements of measurement, survey techniques, and questionnaire design.

- To exploit the depth and richness of currently available victimization data through analytical research on issues of public concern and consequence to the development of national, State, and local criminal justice policy and legislation, with broad dissemination of findings.

- To assist State and local government efforts to improve the administration of criminal justice through (a) promotion of analysis of national data to understand local implications; (b) provision of national guidance on the feasibility, conduct, and utility of local victimization surveys; and (c) provision of a limited set of subnational social indicators derived from the national survey.

- To expand the current victimization survey to include assessment of vulnerability and susceptibility to crime of various segments of the population, and to explore governmental and private approaches for reducing the opportunity for criminal acts and the risk of victimization.

- To examine, through the longitudinal component of the survey, those factors associated with repeated or multiple victimizations to discover appropriate means of reducing such victimizations or minimizing their consequences.

- To use the ongoing national survey to obtain additional information on crime and criminal justice issues through supplemental questionnaires.

The consortium tasked with redesigning the National Crime Survey in the mid-1980s reasoned that "the process of constructing an efficient sample design begins with a detailed description of survey objectives and a clear ranking of these objectives." Hence, it developed its own ranked list of objectives (highest to lowest) and considered costs and benefits of proposed changes relative to each of these objectives (Biderman et al., 1986:19–20):

Recommended Ranking of NCS Objectives

1.0 To provide data that will serve as a set of continuous and comparable social indicators of trends in the rates of victimization by selected crimes of violence and crimes of theft and of factors related to crime and victimization in support of national criminal justice policy, and informed public decision.

2.0 To provide policymakers at the national and state and local levels as well as the research community, with a database concerning crime victims and victimization.

   2.1 To provide empirical information concerning the characteristics of victims and consequences of the victimization that will be useful in designing, implementing, and maintaining victim assistance programs.

   2.2 To provide empirical information on perceived satisfaction with the criminal justice system.

3.0 To facilitate analytical research on issues of public concern and of consequence to the development of national, state, and local criminal justice policy.

   3.1 To provide empirical information that assists individuals and households in avoiding victimization.

   3.2 To provide empirical information relevant to understanding the difference between the reported crime rate (UCR) and the victimization rate.

4.0 To provide, in addition to national data, crime indicators for selected cities and states.

   4.1 To assist state and local governments in evaluating the feasibility and utility of local victimization surveys.

5.0 To gather information on a regular basis concerning attitudes toward crime, criminals, and crime control.

Most of these—notably excluding the provision of "crime indicators for selected cities and states"—were retained in a 1989 listing of "current BJS objectives for the NCS program" (Bureau of Justice Statistics, 1989:5–6).

Current documentation for the public-use NCVS files archived by the Interuniversity Consortium for Political and Social Research (Bureau of Justice Statistics, 2007a:5) provides a concise summary of historical goal statements: "The NCS and NCVS were both designed with four primary objectives: 1) to develop detailed information about the victims and consequences of crime; 2) to estimate the numbers and types of crimes not reported to police; 3) to provide uniform measures of selected types of crimes; and 4) to permit comparisons over time and types of areas." Likewise, the current Census Bureau manual for NCVS interviewers (U.S. Census Bureau, 2003:A1-2) simplifies the discussion to a "primary purpose" and a "secondary purpose":

[The primary purpose is] to obtain, from respondents who are 12 years of age and older, an accurate and up-to-date measure of the amount and kinds of crime committed during a specific six-month reference period.[3] . . . The NCVS also collects detailed information about specific incidents of criminal victimization that the respondent reports for the six-month reference period. [The secondary purpose is that] NCVS also serves as a vehicle for obtaining supplemental data on crime and the criminal justice system; . . . this supplemental information is collected periodically, along with the standard NCVS data.

2-B   DATA ON CRIMES AND VICTIMS

At the most fundamental level, the NCVS is intended to do what the name implies: provide measures of criminal victimization. This statement is a great understatement, though, as the implementation of the NCVS opened up several dimensions of understanding of crime that were not well or systematically considered in earlier measures:

- *Unreported crime:* Before the first crime surveys, no one had an inkling of the magnitude of the "dark figure" of unreported crime, nor was there any systematic knowledge of the factors that facilitated or discouraged victim reporting. The NCVS examines whether victimizations were reported to police and the reasons for reporting or not reporting. See Section 3–F for further discussion.

- *Characteristics of victims:* Before the NCVS, little was known about crime victims except for sketchy profiles (typically of their age, sex, and race) that could be extracted with some difficulty from police case files. There was great interest in learning more about them. At the time the NCVS was being developed, victims were frequently regarded as the "forgotten participants" in the criminal justice system. There was concern at the time about apparently high rates of victimization of the elderly and about domestic violence. The survey promised to shed new light on the victims of different kinds of crime, including a range of victim attributes (such as household income or education) that go unrecorded by the police. See, for example, Klaus and Rennison (2002) and Bachman et al. (1998) on differential patterns in victimization by age; Smith (1987), Smith and Kuchta (1993), Lauritsen and Schaum (2004) on victimization of women; Dugan and Apel (2003) on combined effects of race, ethnicity, and gender; and Levitt (1999) and Thacher (2004) on difference in victimization by income.

---

[3]The wording of this objective is somewhat odd, as it suggests to interviewers that the goal of the survey is to elicit confessions (i.e., "crimes committed") from respondents.

- *Consequences for victims:* Detailed questions on the NCVS yielded the first national picture of the consequences of crime for victims, including extent of injury, hospital stays and insurance coverage, out-of-pocket medical expenses, the value of lost time from work, and other costs. The resulting national estimates have been used in cost-benefit analyses of crime prevention programs (for a review, see Cohen, 2000), and they make the NCVS the only comprehensive national source of information on the costs of crime to victims (see Section 3–E.1).

- *Circumstances surrounding victimization:* The survey includes questions about the location of each incident (at home, in commercial places, at school, etc.); whether household members or other people were present at the scene and if they took any action; and if respondents took any self-protective measures that appear to have improved their situation. Other details include gun and weapon use, whether the crime was an attempted or completed incident, and the number of offenders. See, for example, Rand (1994) and Zawitz (1995) for NCVS findings regarding the use of firearms in crime and Bachman et al. (2002) on the use and effective of self-protective behaviors by victims.

- *Profiles of offenders:* In surveys, victims can be asked to describe offenders and what could be discerned concerning their motivations. For example, among those victims who provided information about the offender's use of alcohol, about 30 percent of the victimizations involved an offender who had been drinking (http://www.ojp.usdoj.gov/bjs/cvict_c.htm; Greenfeld and Henneberg, 2001). The NCVS also includes questions about the role of apparent bias in the targeting of victims, to probe the prevalence of hate crimes (as we discuss further in Section 2–F).

2–C NATIONAL BENCHMARK: NCVS AND THE UNIFORM CRIME REPORTS

Along with the basic goal of providing information on the characteristics of victims, the role of measurements from a victimization survey as a counterpart to official reports to police—shedding light on the so-called dark figure of unreported crime—is an original, long-standing theme. The role of NCVS as a "continuous and comparable national social indication" to the UCR program's official counts is directly envisioned by the BJS authorizing legislation, as discussed in Section 2–F. The duality between the NCVS and the UCR is a major part of the public's awareness of the victimization survey: each year, the release of new data from the NCVS and UCR programs is met with a flurry of news stories on possible upticks and downticks in crime. But the duality may also contribute to confusion and uncertainty among the public and stakeholders like Congress, when duality and complementarity are mistaken for redundancy and it is assumed that the two programs measure exactly the same thing. The conundrum is that the potential for confusion is heightened on those occasions when the NCVS and the UCR suggest trends that go in opposite directions, while the potential for perceived redundancy and waste is heightened when the two measures suggest similar trends.

Given this chapter's focus on historical goals of the NCVS, this section deals with only a limited set of issues: basic definitional differences between the two programs and the national role of the NCVS as a counterpart indicator to the UCR results. Chapter 3 examines broader dimensions of the duality between the NCVS and the UCR, including the tracking of trends in the two series over time (convergence and divergence relative to each other) and the relative quality of the series. Technical details about the UCR are included for reference in Appendix D.

When UCR collection began in 1929, the program was designed to collect information "on a subset of crimes that the International Association of Chiefs of Police (IACP) at the time considered prevalent, serious, and well reported to police" (Cantor and Lynch, 2000:86–87). That core set of crime types—the so-called index crimes of criminal homicide, forcible rape, aggravated assault, robbery, burglary, larceny-theft, and motor vehicle theft—has remained the same over time. Indeed, as detailed in Appendix D, only the crime of arson has subsequently been elevated to the UCR's top tier of measured offenses.

As a survey, the NCVS offered (and still does) a basis for collecting information on more and varied types of victimization incidents than police records permit. Yet the NCVS remains decidedly tethered to some features and concepts of the longer established UCR program. Cantor and Lynch (2000:107) argue that the initial design of the National Crime Survey deviated from the recommendations of the President's Commission on Law Enforcement and Administration of Justice (1967) that led to its founding due to "the attempt to mimic UCR":

> Specifically, the principle of facilitating recall and reporting was compromised somewhat in favor of some of the legal principles and the desire to classify crimes neatly. . . . [The new NCS, administered by the Census Bureau,] restricted its screening to Part I crimes in UCR, such that questions were asked with the intent of eliciting mentions of these crimes and only these crimes. Although the Census instrumentation separated the screening task from the provision of detailed information for classification, there was a one-to-one correspondence between the screen questions and the UCR crimes.

Although the NCVS may have echoed UCR structures in order to establish an equal footing, it remains true that the NCVS is designed to produce that which the UCR "has never attempted to produce: a count of crime that includes serious offenses, like rape, that may never be reported to police" (Robinson, 2007). In providing these data "in a reliable and consistent fashion," Robinson (2007) argues that BJS' sponsorship of the NCVS fills "a distinctly federal role," generating data that no single state can afford to produce on a regular basis.[4] We discuss victimization surveys that have been conducted by individual states in Section 3–D.

2–D ANALYTIC FLEXIBILITY

In the early 1970s, as now, the UCR gathered monthly and yearly crime totals in only a few (currently eight) broad categories, and the FBI received them only at the jurisdiction level. One promise of the NCVS was that the incident-level data it produced could be used to address a variety of research and policy questions, because of the analytic flexibility of surveys. One feature of this flexibility is that the NCVS can be analyzed at multiple levels. The survey gathers reports of individual and household victimization, and most descriptive publications examine rates of crime at those levels. However, the data can be organized in a variety of ways to address descriptive and analytic questions.

- *Households "touched by crime"*: BJS reports have combined data for households and all of the individuals living in them, to characterize the percentage of households that have had some recent experience with crime (e.g., Klaus, 2007). Families are another analytic unit that can be distinguished in the survey, and a variety of family crimes are within the scope of the survey (Durose et al., 2005).

- *Crimes by location*: Fairly detailed descriptions of the location of incidents are gathered in the NCVS, revealing that 22 percent of the victims of violence were involved in some form of leisure activity away from home at the time of their victimization; 22 percent reported they were at home, and another 20 percent mentioned they were at work or traveling to or from work when the crime occurred. See, for example, Warchol (1998) for specific study of workplace violence. Schools are another important locus for crime problems, and the survey has been used extensively to examine school crime (as described further in the next section).

---

[4]Robinson (2007), testifying before a congressional appropriations subcommittee, argued that NCVS should be afforded "a *broadened role* in helping in our understanding of victimization. BJS should be provided with increased funding to enable it to measure crime on a *state-by-state basis*, and even to the level of large cities."

- *Neighborhood effects on victimization:* For special analyses, data for the census tracts in which NCVS respondents live have been combined with their responses to the survey, to examine the effects of such factors as concentrated poverty and residential instability on victimization rates and the characteristics of crimes (Baumer, 2002; Baumer et al., 2003).

- *Metropolitan area rates and trends:* The sample design of the NCVS is capable of providing estimates of crime for the largest metropolitan areas. NCVS data for some of the largest areas have been used to describe local area victimization trends and to compare these to trends in the UCR for the same areas (Lauritsen and Schaum, 2005; Lauritsen, 2006a).

- *Long-term trends:* With the exception of the redesign in 1992, the methodology of the NCVS has remained fairly consistent over time. The data have been used to track long-term national trends in crime as well as long-term victimization trends for different subgroups, such as males and females, blacks and whites, and different age groups (see, e.g., Steffensmeier and Harer, 1999; Klaus and Rennison, 2002).

- *Longitudinal patterns:* Because NCVS respondents are interviewed up to seven times while they are participants in the crime panel, data can be organized to trace patterns of victimization over time for individuals and households that do not change residences. These data have been used to study such issues as the effect of reporting crime to the police on subsequent victimization and the effect of victimization on the decision to move (see, e.g., Conaway and Lohr, 1994; Dugan, 1999).

2–E   TOPICAL FLEXIBILITY: NCVS SUPPLEMENTS

From the beginning it was anticipated that the NCVS would provide a flexible vehicle for gathering occasional or one-time data to supplement the ongoing core data required to track national trends in crime. These supplements might be supported by research grants, allocations by Congress, or by contributions by partner agencies who wished to gather specialized data relevant to their responsibilities. One responsibility is to design ways to fold these requests into ongoing data collection in ways that do not disrupt the continuing flow of data. The supplements have made significant contributions to research and policy. They include:

- *Crime seriousness:* The first NCVS supplement gathered national data on the perceived seriousness of crime, information that has been used to differentially weight incidents to reflect their impact on the public. Subsequent BJS analyses have used both NCVS data and information from the American Housing Survey to describe perceptions of crime severity (DeFrances and Smith, 1994); on a one-shot basis, BJS also collaborated with the Office of Community Oriented Policing Services in conducting community safety surveys by telephone in 12 cities, wholly distinct from the NCVS (Smith et al., 1999).

- *Attitudes and lifestyles:* Another supplement gathered extensive data on the attitudes of individuals and the relationship between crime and how they conduct their lives (Murphy, 1976; Cowan et al., 1984).

- *Crime in schools:* BJS, in collaboration with the National Center for Education Statistics, periodically collects data on aspects of school crime through the School Crime Supplement to the NCVS. Respondents age 12 and older attending school are asked about their school environment. Information is gathered on the availability of drugs at school, the existence of street gangs and the prevalence of gang fights, the presence of guns at school, victimizations, and their fear of being attacked or harmed. Results from the various administrations of the School Crime Supplement, as well as related data resources, are described by Dinkes et al. (2007).

Another important supplement, the Police-Public Contact Survey, is described in the next section as an example of a legislative mandate.

2–F   LEGAL MANDATES

Finally, in outlining the goals and objectives of the NCVS, it is important to consider responsibilities that the survey (and BJS) must accomplish due to requirements in law.

The duties and functions of BJS laid out in its enabling statute [42 U.S.C. 3732(c)(2)–(3), originally enacted as P.L. 90-351, the Omnibus Crime Control and Safe Streets Act of 1968] directly authorizes the bureau to "collect and analyze information concerning criminal victimization, including crimes against the elderly, and civil disputes." The authorizing language further directs that BJS should:

> collect and analyze data that will serve as a continuous and comparable national social indication of the prevalence, incidence, rates, extent, distribution, and attributes of crime, juvenile delinquency, civil disputes, and other statistical factors related to crime, civil disputes, and juvenile delinquency, in support of national, State, and local justice policy and decisionmaking.

Although the use of the resulting data for developing national policy is noted in the language, the preamble to BJS' authorization clearly hearkens to the agency's origins as part of the Law Enforcement Assistance Administration: in its work, BJS "shall give primary emphasis to the problems of State and local justice systems" (42 U.S.C. 3731).

Over the years, Congress has mandated that specific information related to criminal victimization be collected by BJS. Some of these mandates explicitly direct that these new data collections be added to the NCVS, while others become linked to the NCVS because the insertion of a supplement to the survey is seen as the most expedient solution. An example of a direct-to-NCVS mandate is a provision in P.L. 106-534, the Protecting Seniors from Fraud Act of 2000. As part of a broader study of the prevalence of crimes against seniors, section 6 of the act requires BJS, "as part of each National Crime Victimization Survey," to:

> include statistics relating to—
>
> (1) crimes targeting or disproportionately affecting seniors;
> (2) crime risk factors for seniors, including the times and locations at which crimes victimizing seniors are most likely to occur; and
> (3) specific characteristics of the victims of crimes who are seniors, including age, gender, race or ethnicity, and socioeconomic status.

Response to this legislative mandate has focused on the crimes of identity theft, credit card fraud, and bank fraud.

Similar language in P.L. 105-301 directed that the NCVS include measures of "the nature of crimes against individuals with developmental disabilities" and "the specific characteristics of the victims of those crimes." This act, the Crime Victims with Disabilities Awareness Act of 1998, led to implementation (after pilot testing) of NCVS questions gauging whether victims of crime were in poor health, had any physical or mental impairments, or had disabilities that affected their everyday life. They are also asked to judge if any of these provided an opportunity for their victimization.

The Violent Crime Control and Law Enforcement Act of 1994 (P.L. 103-322, Section 210402) included a brief provision that "the Attorney General shall, through appropriate means, acquire data about the use of excessive force by law enforcement officers." As a direct result of this mandate, BJS developed the Police-Public Contact Survey to measure the extent of all types of interactions between the police and members of the public (of which those involving "excessive force" is logically a subset). The survey was first conducted on a pilot basis in 1996 (Greenfeld et al., 1997); after refinement, it was fully fielded as an NCVS supplement in 1999 and has become a continuing occasional supplement. The survey gathers detailed information about the nature of police-citizen contacts, respondent reports of police use of force and their assessments of that force, and self-reports of provocative actions that they may have themselves initiated during the encounter. Data from the supplement have been used to in analyses of possible levels of racial profiling by, for example, Engel and Calnon (2004) and Engel (2005).

Other examples of legislative mandates impacting the NCVS include:

- The Hate Crime Statistics Act of 1990 (P.L. 101-275), which directed the Attorney General to "acquire data, for each calendar year, about crimes that manifest evidence of prejudice based on race, religion, disability, sexual orientation, or ethnicity." In this case, the NCVS was not explicitly mentioned as the data collection vehicle. In response to the mandate, the NCVS now includes questions probing the possible role of prejudice or bigotry in motivating offenders. For results from the hate crime questions on the NCVS, see Harlow (2005), and see Lee et al. (1999) Lauritsen (2005) for additional discussion of the development of the questions. We discuss the addition of hate crime questions further in Section 3–C.1.

- Family violence reporting provisions in P.L. 100-690, which directed BJS, "through the annual National Crime Survey, [to] collect and publish data that more accurately measures the extent of domestic violence in America, especially the physical and sexual abuse of children and the elderly."

The addition of NCVS content and topics due to legislative mandates implicitly recognizes the importance of the NCVS as a source of information. However, it also adds to the burden on the survey and on BJS as a whole, particularly since these changes have not been matched by budgetary increases.

– 3 –

Current Demands and Constraints on the National Crime Victimization Survey

THIS CHAPTER BUILDS ON THE DISCUSSION of historical goals in Chapter 2 by examining some contemporary issues and challenges facing the measurement of victimization, in particular the demands and constraints placed on the National Crime Victimization Survey (NCVS). We begin in Section 3–A by discussing survey nonresponse, an emerging challenge facing modern surveys of all types, including federal surveys like the NCVS. Section 3–B discusses basic challenges of self response in measuring victimization, including discussion of crimes that are not well measured in police reports and that are inherently hard to measure: the capability of the NCVS to provide information on these is at once a great strength of the survey and a major, ongoing technical challenge. Section 3–C expands the discussion of analytical flexibility from the previous chapter to include issues of flexibility in measuring new types of victimization as well as changes in basic methodology, including subnational estimation, to meet user needs. We then turn to a basic underlying question—What is the value of measuring victimization?—and consider how the cost of the NCVS compares with various benchmarks in Section 3–E. Section 3–F turns to basic issues related to the coexistence of two related measures of crime in the NCVS and the Uniform Crime Reports (UCR): the general correspondence of the two series over time and resulting questions about the need for two independent measures. We conclude in Section 3–G, considering both the historical goals of the NCVS (Chapter 2) and the challenges described in this chapter to assess the basic utility of the NCVS.

3–A   CHALLENGES TO SURVEYS OF THE AMERICAN PUBLIC

3–A.1   The Decline in Response Rates

With a response rate of 91 percent among eligible households (84 percent of eligible persons) as of 2005, the NCVS enjoys response and participation rates that are highly desirable relative to other victimization and social surveys. However, the NCVS response rates have declined over the past decade; in 1996, the NCVS household- and person-level response rates were 93 and 91 percent, respectively (Bureau of Justice Statistics, 2006a). Figure 3-1 illustrates the recent growth in the noninterview rate in the NCVS and one component of that rate in particular: refusals by anyone in the contacted household to participate in the survey. The figure presents these noninterview and refusal rates for both initial contacts (interview 1, conducted by personal visit) and for all data collection in the year (including telephone and personal interviews for contacts 2–7 with sample addresses); the initial and aggregate rates generally track each other closely.

The decline in response rates is a situation faced by almost all household surveys in the United States (Groves et al., 2002). For instance, the General Social Survey, a cross-sectional household survey conducted by the National Opinion Research Center at the University of Chicago, has experienced a declining response rate in recent years, from response rates in the high 70s (percentages) and a peak of 82 percent in 1993 to 70–71 percent in 2000–2006.[1] Remedies to address declines in response rate continue to be developed. The Substance Abuse and Mental Health Services Administration, contracting with the Research Triangle Institute, implemented a $30 incentive in 2002 in order to induce respondents to return questionnaires for the National Survey of Drug Use and Health due to declining response rates. The highly detailed National Health and Nutrition Examination Survey, conducted by Westat, has experienced a similar reduction in response.

There is little evidence that the loss of response rate over time is primarily a function of what organization conducts the survey: many of the federal surveys collected by the U.S. Census Bureau (including the NCVS) for various sponsors have shown declines as well. Atrostic et al. (2001) describe measures of nonresponse for six federal surveys (including the NCVS) between 1990 and 1999, documenting consistent declines in response; Bates (2006) updates the series through 2005. An example cited in those works

---

[1]See http://www.norc.org/Projects+2/GSS+Facts.htm [8/20/07].

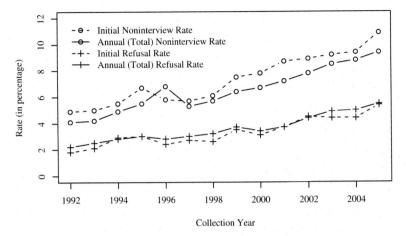

Figure 3-1 Noninterview and refusal rates, National Crime Victimization Survey, 1992–2005

NOTE: Refusal rate is defined as the number of eligible interviewing units not interviewed because occupants refused to participate divided by the total number of eligible interviewing units. Refusals are a component in the noninterview rate, which also includes interviews not completed due to other reasons (e.g., language difficulty or no one at home). Noninterviews are termed "Type A" results. Rates are based on unweighted data.

SOURCES: Data from Bates (2006); definitions from Atrostic et al. (2001).

is the Consumer Expenditure Diary (CED) survey, data from which are an input used to derive the consumer price index. Between 1991 and 2003, the initial nonresponse rate (failure to respond to the first interview, which, like the NCVS, must be done by personal visit) steadily increased from about 15 percent to about 30 percent; in 2005, the CED had an overall nonresponse rate of 31.1 percent (Bureau of Labor Statistics, 2007:87).

A notable exception to the pattern of declining response rates in federal surveys is the American Community Survey (ACS), the replacement for the traditional decennial census long-form sample that asked census respondents for additional social and demographic information. However, the ACS also holds a distinct advantage over other Census Bureau surveys because—inheriting from the decennial census—responses to the ACS are required by law (and respondents are so advised). A test conducted by fielding the ACS with wording on the mailing materials suggesting that response is voluntary (e.g., "Your Response Is Important to Your Community") rather than mandatory (e.g., "Your Response Is Required by Law") demonstrated a radically reduced mail response rate: an overall drop of 20.7 percentage points (Griffin et al., 2003, 2004).

The fact that declines in response rates are not isolated to private surveys rather than federal surveys (or vice versa) suggests that large-scale changes in the relationship between survey data collectors (generally) and the U.S. public have occurred in recent years. While there is no convincing empirical evidence to test alternative theories of the causes of the decline, the most popular hypotheses include:

- the lack of trust in the institutions requesting the survey participation;
- confusion in potential respondents' minds between marketing approaches and survey participation requests;
- loss in discretionary time at home due to increased commute time to work and other out-of-home activities;
- increase in the sheer volume of survey participation requests, making participation in any particular survey less novel; and
- the increased investment in electronic and other devices to prevent strangers from contacting the public.

There are two principal forms of nonresponse, each of which appears to have separate causes. The first is unit nonresponse: for a household survey like the NCVS, this is nonresponse that arises because the household at a particular address could not be contacted or declined to participate at all. The inability to contact U.S. households is driven both by apparent increases in the out-of-home activities of the public and by changes in how the public views approaches from strangers. There are now more walled subdivisions, locked multiunit structures, and intercom systems that permit residents to control the access of strangers to their housing units. For telephone contact, answering machines and "caller ID" features permit residents to limit telephone contact to those persons they wish to talk to. Hence, populations that invest in these housing features and appliances are disproportionately not contacted. These tend to be urban dwellers, younger, more transient persons, and those who live alone. This broader, structural form of nonresponse is inherent to all surveys. As we discuss in Section 5–B, it is an open question whether the administration of the NCVS by the U.S. Census Bureau is a net positive or negative (or neither) in affecting unit response.

The second type of nonresponse is within-unit nonresponse: given that contact is successfully made at an address, do all the survey-eligible persons at that address cooperate and answer the survey questions? There is evidence that persons who are interested inherently in the announced topic of the survey tend to respond (Groves et al., 2004). There is also evidence that women cooperate more prevalently than men (DeMaio, 1980); that urban dwellers cooperate less frequently than those in rural areas (Groves and Couper, 1998); and that those who live alone and middle-aged persons are less cooperative. To the extent that person nonresponse relies on interest in the survey's topic area, the purview of the NCVS presents complications on both ends of a continuum. For people who have been victimized—particularly highly sensitive crimes like sexual assault—interviewers may face a difficult task in building rapport so that respondents are willing to talk about their experiences. Likewise, interviewers have to be trained to handle the opposite situation: people who have not experienced recent victimization and hence attempt to bow out of the survey because they think it irrelevant.

Survey design features may have some role in affecting response rate, or at least in curbing the loss of response rate. There is evidence that longitudinal surveys of persons—in which multiple contacts are made with the same households and people, forging longer term "relationships" between interviewers and subjects—have experienced lower declines in participation. The NCVS is a longitudinal survey of *addresses*, not persons, and thus may be affected by turnover of individual persons or families at sampled addresses. However, nonmovers—people who remain at the same address over time—can experience up to seven NCVS requests and thus—conceptually—the NCVS response rate should enjoy some benefit from those repeated contact efforts. As Lepkowski and Couper (2002) note, however, the propensity to respond in later waves of a longitudinal survey is dependent on the enjoyment of the prior wave. If NCVS respondents in one wave find the survey less than pleasant, there may be lower propensity to respond in the next wave.

Telephone surveys appear to suffer more dramatic nonresponse rate increases than face-to-face surveys. This finding primarily comes from evidence from random-digit-dialed surveys (Curtin et al., 2000). The NCVS does use the telephone for waves 2–7 of interviewing, but, given this is common to other longitudinal surveys, it is unlikely that the use of the telephone in NCVS interviewing is, in itself, a principal cause of lower response rates.

It is important to note that nonresponse rates are only proxy indicators of one aspect of the quality of NCVS estimates. The key issue is whether the propensity to be successfully measured among NCVS sample members is correlated with the likelihood of victimization. Tests conducted alongside the British Crime Survey (BCS) and the Scottish Crime Survey (SCS) provide useful evidence along these lines. Lynn (1997) describes a BCS experiment that urged nonrespondents to provide some limited information; people who said that they did not want to be interviewed were pressed to give very short answers about the extent of recent victimizations against them. These capsule assessments were found to be consistent with victimization estimates among people who completed the survey. Similar findings were registered in an SCS test documented by Hope (2005); that test also compared responses gained by face-to-face interviewing compared with telephone response, since a change to telephone collection was being considered for the 2004 administration of the SCS. In the test, face-to-face and telephone interviews were conducted in parallel; the face-to-face interviews recorded a 67 percent response rate compared with 49 percent by telephone, with the difference attributed to refusals to be interviewed. When victimization estimates from the two modes of administration were compared, the telephone rates were found to be consistently higher, raising the possibility that the telephone administration had the effect of oversampling persons with incidents to report. A follow-up test recontacted some respondents from the first survey and compared victimization estimates for the group of people who responded on the first contact with those who had their refusals "converted" to responses in the second pass. Victimization rates were found to be lower for the "converted" group than the initial respondents, corroborating the hypothesis that refusals are more likely to include nonvictims (people with no incidents to report) rather than victims.

3–A.2   The Rise in Survey Costs

From the fiscal and operational standpoint, the major consequence of increased nonresponse is increased survey costs. These are incurred when the data collection effort seeks to maximize response rates, devoting field resources to repeated attempts to contact households for interviews. Cost inflation would be more modest if only one call were made to each household. In the NCVS and other surveys seeking high-quality estimates, repeated callbacks and efforts at persuasion are introduced on sample cases that have not yet been interviewed. Repeated calls in face-to-face surveys require the interviewer to drive to the sample unit and attempt contact. If no one in the household is at home, another call—often on another trip—is required. If a contact is achieved but the householder is reluctant to participate at the time, another call is made. What results from such a recruitment protocol is that noninterviews require more effort than interviews; the cost of a failure is larger than the cost of a successful interview. As the difficulty of making contact and gaining cooperation increases over time, the costs of the total effort increase if response rates are to be maintained.

In short, attempting to achieve high response rates in a survey of a population presenting growing difficulty in making contact and gaining cooperation will lead to cost inflation.

3–A.3   The Linkage Between Response Rates and Nonresponse Error

It is traditional to attempt to maximize response rates in an effort to reduce nonreponse error. This flows from a simple deterministic view of nonresponse error in a sample mean (like the number of victimizations reported divided by the number of persons) as a function of nonresponse rates and the difference between respondent and nonrespondent means. Increasingly, empirical studies have shown that a stochastic view of nonresponse is more appropriate, viewing each decision to be a respondent as subject to uncertainty. In this view, high correlation between the likelihood of participating and the survey measures produces nonresponse bias in such descriptive statistics. Which NCVS estimates might illustrate such links to response propensities is at this point an unknown question. Some NCVS estimates might be biased from the nonreponse and others might not. New studies are appropriate to gauge what value BJS should place on high NCVS response rates, given both the current design and for future alternative designs.

Given the ubiquity of the nonresponse problem across federal surveys, recent U.S. Office of Management and Budget (2006b) guidelines call for analyses of nonresponse bias when either unit or item nonresponse hits certain levels. The NCVS unit response rate is such that this threshold has not been crossed; still, we know of no effort by BJS or the Census Bureau to mount a full nonresponse bias study for the NCVS.

3–B CHALLENGES OF SELF-RESPONSE IN MEASURING VICTIMIZATION

3–B.1 Cognitive Challenges: Telescoping and Forgetting

As noted above, the NCVS emerged from the National Crime Survey only after several years of conceptual development and methodological research. The research was path-breaking in that it helped launch what is sometimes called the cognitive aspects of survey measurement (CASM) movement (aided by a Committee on National Statistics workshop, National Research Council, 1984). Much of the labor of that redesign effort (Biderman et al., 1986) targeted improved reporting among respondents to the NCS. Importing key notions from cognitive processing models, it was noted that autobiographical reports were fraught with weaknesses. Memories were viewed as being formed at an "encoding" step, in which sense-based observations were retained, often in a manner that was heavily dependent on the situation during the experience of the events. Not all encoded memories were easily retrieved upon a desire to do so. The studies found that "forgetting" events that did occur was a challenge to the survey. Consistent with long-standing results from cognitive psychology it was found that events that did not induce emotional reactions ("nonsalient"), those that happened frequently, and those that occurred far back in time tended to be underreported. Thus, "forgetting" was a problem for the NCS.

Research on context-dependent recall suggested that individual words and mentions of types of related events were effective "cues" to memory recall. Much of the research on the screener questions, therefore, was attempting to improve the rate of reporting of incidents as a way to attack the problem of "forgetting." The "short cue" version of the instrument that resulted from the research attempted to provide a rich set of cues for each victimization type. In this regard, a marriage between the incident report and the screening questions was key. The screener questions were designed to maximize recall, even at the risk of overreporting incidents through duplicate reports about the same event or misdating of an event that occurred outside the reference period. The role of the incident reports was to duplicate those reports.

The finding that forgetting was a function of the salience of the event to the person and the length of time since the event implied that smaller victimizations occurring further back in time were most fraught with reporting errors. The length of the reference period (the time from the start of the eligible time period for events to be in scope to the end of the period) and the length of the recall period (the time between the start of the in-scope period and the day of the interview) were issues that could affect the quality of reports. Longer periods yielded poorer reports (Miller and Groves, 1985; Czaja et al., 1994), generally a mix of forgetting and misdating events. The redesign recommended a 6-month reference period, a recommendation based on the findings of increased measurement error due to forgetting and telescoping in 12-month reference periods.

There was another antidote to misdating or telescoping errors, which was already in place in the NCS—the use of a bounding interview. A bounding interview in the context of the NCS was the first wave interview with each respondent, in which events in the 6-month reference period before the interview were reported. No data from the bounding interview were used in estimation (another recommendation stemming from findings of forward telescoping errors). Instead, the events reported in the bounding interview were made known to the second wave interviewer to verify that a incident reported in that interview was not a duplicate of a report in the first, bounding interview. This was thought to reduce forward telescoping errors in the NCS estimates. Some research in the redesign focused on whether the data from the bounding interview might be integrated through statistical models into the estimates, but that never led to such a recommendation.

The panel notes that the design features of the reference period, the cuing mechanisms of the screener questions, the nature of the incident reports, and the use of the bounding interview technique are mutually connected. It is difficult to evaluate one of these features without simultaneously considering the others.

3–B.2　Measuring Hard-to-Measure Crimes

A key conceptual strength of the NCVS is its ability to elicit information about victimization incidents that are not reported to police. This is particularly the case for such personally sensitive crimes as rape and domestic violence, as well as simple assault and other incidents that are acts of violence but that victims may decline to report (or judge that they are not sufficiently severe to report) to authorities. The 1992 redesign concentrated on improving the screening questionnaire to more effectively and accurately probe respondents to recall and report such incidents, doing so by increasing the density of cues and using multiple frames of reference. Both the pre- and postredesign questionnaires emphasized the need to broach questions in language that is accessible and understandable (and not steeped in legal jargon) in order to boost cooperation and accurate recall.

However, improvements to the screening and cuing procedures leave open the question of whether reporting of hard-to-measure crimes is full and complete and whether other approaches may be preferable. Conceptually, the measurement of sensitive crimes through personal interviewing is a sounder approach than reliance on police reports, but it is important to consider that some crimes that are not reported to police may not be reported to interviewers, either. From the technical perspective, some hard-to-measure crimes—notably domestic violence—present continuing measurement challenges due to their high frequency; determining an accurate count is a formidable difficulty, and detailed information on specific incidents even more so. In this section, we briefly discuss the challenges in getting accurate survey reports in two areas: measurement of rape and domestic violence and description of repeated (series) victimizations.

Rape, Domestic Violence, and Simple Assault

Several researchers have reported that the NCVS yielded lower estimates of the incidence of rape and domestic violence than other surveys. For example, before the 1992 switch to a redesigned instrument, the National Crime Survey produced estimates of domestic violence that were an order of magnitude smaller than those produced by other surveys.[2] Similar results obtain

---

[2]Bachman and Taylor (1994) compared the then-available estimates of family violence against women from the NCVS to results from the National Family Violence Survey (NFVS). The cross-sectional NFVS was conducted twice, in 1975 and 1985, by the Family Research Laboratory of the University of New Hampshire and reached a sample of 2,143 and 6,002 households in the two administrations, respectively. The survey suggested that around 160 per 1,000 married couples experienced at least one "violent incident" in 1975 and 1985. By comparison, the NCVS—which did not ask specifically about violence by family members before the redesign—yields an estimate of the annual rate of famly violence against women of just 3.2 per 1,000. However, the two survey estimates are not directly comparable because they frame

Table 3-1 Rape and Assault Rates, National Crime
Victimization Survey and National Violence
Against Women Survey, 1995

	NVAWS	NCVS	Adjusted NCVS
Rape	8.7	1.9	2.6
Intimate Partner Assault	44.2	6.6	26.7
Assault	58.9	25.8	80.4

NOTES: Rates per 1,000 population. "Adjusted NCVS" are
estimates calculated by including the reported count of incidents in
series victimizations in the estimates.

SOURCE: Rand and Rennison (2005).

for rape: the pre-1992 NCS instrument did not define rape for the respondent and did not directly ask respondents whether they had been victims of attempted or completed rape (Bachman and Taylor, 1994:506). The redesigned NCVS asked respondents more directly about family violence and rape. The new instrument asks specifically about violence or threats perpetrated by "a relative or family member." The redesigned instrument also asks more directly about "unwanted sexual activity," including from those who are well known to respondents. A range of probes also distinguishes between verbal and other threats, attempts, and completed rapes.

Post-redesign research comparing results from the new NCVS instrument with previous versions has largely focused on broader crime categories (Kindermann et al., 1997) or differences by analytic groups (Cantor and Lynch, 2005) and not on specific crimes like rape or domestic violence. Still, several studies compared NCVS rates with those generated by other surveys on these categories. Rand and Rennison (2005) contrasted rape and assault rates in the 1995 NCVS with those from the National Violence Against Women Survey (NVAWS), a telephone survey of U.S. adults. They obtained the estimates for annual incidence shown in the NCVS and NVAWS columns of Table 3-1.

Rand and Rennison (2005) suggest several explanations for the discrepancy between the data sources: the NVAWS may elicit more victimizations by asking about rapes and assaults more explicitly: the NVAWS may be more vulnerable to telescoping, in which incidents outside the reference period are included; or the two data sources may diverge because of their measurement of recurring victimization. As a one-time, single-interview survey, the NVAWS had no capacity for bounding responses, "suggesting that [NVAWS]

---

incidents differently: the NFVS instrument asked specifically about "conflict" among family members rather than more detailed probes about violent incidents.

estimates are likely inflated to some unknown degree" (Rand and Rennison, 2005:274). (Response rates in the NVAWS are also much lower than the NCVS, although it is difficult to know what biases might result.) The NCVS records as a single series victimization a group of six or more victimizations that were similar in nature but difficult for the respondent to recall individually. Rand and Rennison (2005:275) estimate that series victimizations account for about 10 percent of violent incidents against women. BJS publications exclude series victimizations from annual estimates. After adjusting for age and crime types and counting the number of incidents among series victimizations, Rand and Rennison (2005) obtained the estimated rates of annual incidence reported in the "Adjusted NCVS" column of Table 3-1.

Despite the adjustment for series victimization, rates of rape and intimate partner assault are lower in the NCVS than in the NVAWS. However, Rand and Rennison (2005:279) found that the difference is statistically significant only in the case of intimate partner assault. The discrepancy between the data sources is largest for intimate partner violence, suggesting that at least part of the divergence may be due to the classification of intimate partners rather than the measurement of victimization.

Research on the NCVS redesign also suggests that measurement of simple assaults (without a weapon resulting in minor injury) also depends closely on the survey instrument. With a broader screening interview that cued respondents to consider events they might not define as crimes, the redesigned survey recorded roughly twice the number of simple assaults than the old NCVS (Lynch, 2002). While it is difficult to gauge whether there is still underreporting of less serious personal crime in the NCVS, research on the redesign underlines the sensitivity of estimates to the survey instrument.

The possibility that such crime types as sexual victimization and domestic violence may still be underreported in the standard personal interview context—despite improvements in cuing and screening—highlights the importance of researching means for incorporating self-response options in the NCVS. These include such approaches as web administration and turning the computer laptop around for parts of an interview so that respondents read and answer some questions without interaction with the interviewer. We discuss these further in Chapter 4.

Repeated Victimizations

Repeated victimizations may be underestimated in the NCVS because of the way in which series victimizations are handled. As described in Section C–3.d, NCVS interviewers collect specific information (using an Incident Report form) for each victimization incident reported by a respondent except in instances when six or more very similar incidents occurred within the 6-month reference period. In those cases, a single incident form is completed based on the details of the most recent incident. BJS excludes these series victimizations from its standard NCVS estimates, although basic counts of series and nonseries victimizations are tabulated (see, e.g., Bureau of Justice Statistics, 2006a:Table 110).

Prior to the NCVS redesign in 1992, the threshold for defining a series victimization was three or more similar incidents. The change in threshold provides for somewhat fuller accounting of crime types in which repeated victimization may occur; as in the previous section, domestic violence and intimate partner violence are examples in which this may apply. We know of no research that has estimated the effect of the redesign on the reporting of series victimizations—that is, over and above the emphasis on more effective screening and elicitation of incidents, whether the NCVS instrument is more likely to generate reports of crimes for which series victimization rules would apply. Still, the manner in which series victimizations are collected and counted is an important methodological concern, one that leads to concern about whether some crimes are underestimated as a result.

The Rand and Rennison (2005) results indicate that individually counting series victimizations can help bring the NCVS more into line with other surveys. The scope and effect of series victimizations are also analyzed by Lynch et al. (1998) and Planty (2007); Planty and Strom (2007) compare the effects of different counting rules with resulting instability in the estimates.

Some have used the panel design of the NCVS to estimate repeat victimization. This is a difficult analysis because residential mobility contributes to attrition from the panel, and victimization contributes to residential mobility. Naive panel estimates may underestimate repeat victimization because they undercount victimization of those who have moved and been lost from the survey. Ybarra and Lohr (2002) impute victimization rates to respondents who are lost to residential mobility. They obtain very high repeated rates for violent crime and domestic violence, but these estimates are highly sensitive to the missing data model.

3–C FLEXIBILITY IN CONTENT AND METHODOLOGY

In Section 2–D we discussed the long-standing goal of analytic flexibility of the NCVS, being able to accommodate different types of products. In this section, we expand the discussion of flexibility to include emerging issues in topic areas covered by the NCVS (principally through the use of supplements) and in general methodology.

3–C.1 Is the NCVS Flexible Regarding Changes in Victimization?

As a survey-based method of data collection, the NCVS has the capacity to be a relatively timely and flexible instrument for gathering information about "new" types of crime that are of concern to the public. Since its inception, the NCVS survey instrument has added new measures of criminal victimization and improved existing measures; this was particularly the case with the 1992 redesign, which was intended to improve the survey's measures of rape and sexual assault, nonstranger violence, and other "gray area" victimizations. However, the most common option to provide flexibility in topical coverage in the NCVS has been through the addition of supplemental questionnaires, most often at the behest of other government agencies. School violence is one example of a type of victimization for which periodic supplements to the NCVS have been developed and administered, in this case with the cooperation and sponsorship of the National Center for Education Statistics. Conducted in 1999, 2002, and 2005, the School Crime Supplement provides estimates of crime independent of the statistics gathered by police or by the schools. Over time, some of these supplemental questions have migrated into the main NCVS content, as with questions related to hate crimes.

In theory, a survey is a relatively nimble data collection vehicle—certainly compared with official-records methodology, in which changes in data collection depend on the cooperation of the myriad local agencies that assemble raw data—and so the NCVS instrument (or individual modules) should be able to be rapidly moved from concept to data collection. In practice, however, this process has often taken quite considerable amounts of time. For instance, the measurement of hate crimes using the NCVS began in response to a White House announcement in 1997 that directly offered the NCVS as the instrument of choice for estimating this crime. Research and development of questions using multiple rounds of focus groups and cognitive testing began soon thereafter. Nonetheless, the final set of questions was not administered to the full sample until 2000 (Lee et al., 1999; Lewis, 2002; Lauritsen, 2005). Some of the delay resulted from the complexity of the issue: for example, some focus group participants had trouble deciphering the hate crime terminology, others were unclear about the kinds of evidence that were necessary for such a designation, and some felt that queries about sexual orientation should not be asked. Still other factors that contributed to the delay resulted from the fact that the survey had not yet been fully computer-automated because of persistent budget difficulties.

To some extent, the perceived slowness in implementing new measures and rigidity in approach have been attributed to the Census Bureau as the data collector for the NCVS and other federal surveys. Certainly, major change does not occur easily or quickly in the bureau's flagship product, the decennial census—for instance, the switch to the mail (rather than personal visit) as the principal collection mode for the 1970 census was preceded by major tests dating back to 1948. More recently, the 2006 full-scale implementation of the bureau's American Community Survey followed a decade of pilot testing and a midscale implementation as an experiment in the 2000 census (National Research Council, 2006). With specific regard to the NCVS and other demographic surveys, some delay in fielding changed questions is almost certainly due to what is typically considered a good thing: the Census Bureau's keen attention to cognitive testing in order to try to ensure that questionnaires are clear to respondents. For a survey like the NCVS that asks many questions about hard-to-define (and hard-to-discuss) concepts without seeming legalistic in tone, cognitive tests and other pretesting can be particularly valuable. In addition, some time is required for the U.S. Office of Management and Budget to review, process, and clear proposed survey forms, as they are required to do by law.

In comments to our panel, officials in charge of the British Crime Survey (BCS) noted that they can typically add and change items on the BCS questionnaire within months from the time a decision is made in the United Kingdom's Home Office. By comparison, even though the survey is now fully computerized, Census Bureau representatives noted that a two year lead time should be considered typical. In practical terms, the slowness of the process at the Census Bureau has made the NCVS less flexible than victimization surveys in other countries and, in turn, less responsive to short-term needs for information about victimization and its outcomes. However, the trade-off between rapid turnaround and end data quality is admittedly complex.

Methodological Issues

With the 1992 implementation of the redesign—and a subsequent 14-year transition to all-electronic survey instruments—the NCVS became an important adopter of computer-assisted telephone interviewing (CATI) and computer-assisted personal interviewing (CAPI) methodologies. Although the use of CATI interviewing from centralized sites and the switch from paper questionnaires to CAPI were commonly billed as a major potential source of cost reductions, many survey organizations have found that these steps toward survey automation have fallen short of low-cost promises; see National Research Council (2003b) for a fuller discussion. Indeed, as part of its planned set of cost containment measures for 2007, BJS and the Census Bureau dropped the use of Census Bureau CATI call centers for the NCVS. However—consistent with the NCVS objective of accurate data collection—the strong benefits of computer-based survey techniques must be emphasized. Properly implemented, the question-to-question skip patterns of an electronic survey instrument can make interviewers' tasks easier and quicker and ensure that respondents are guided through portions of the questionnaire (e.g., the screening questions) in a more uniform manner. Electronic administration also permits the use of basic editing routines during the course of the interview, allowing for the correction of contradictory answers and data entry errors. Although centralized CATI implementation has not reduced survey costs as much as hoped—an outcome that is certainly not unique to the NCVS—it is important that the NCVS continue to explore methodological advances that may produce greater accuracy. That said, altering the mix of computer assistance is a change that requires careful pretesting.

By its nature, the NCVS requires respondents to recall and describe events in their past that are unpleasant or uncomfortable at best and intensely traumatic at worst. Hence, a possible methodological improvement suggested by other survey research is incorporating self-response modes to the survey. Computer-assisted self interviewing (CASI) techniques effectively turn around the CAPI dynamic: rather than interviewers reading questions from a computer laptop screen, the laptop is handed over so that only the respondent sees the questions (and his or her answers). A further variant of the basic technique, audio CASI or ACASI, has respondents listen to questions through headphones while going through a questionnaire on the computer-screen. The basic motivation of CASI is that respondents may be more likely to divulge socially sensitive information if they can do so with privacy and without verbally reporting to an interviewer. ACASI research suggests that the methodology is effective in eliciting more reports of sensitive information than standard interviewer-administered approaches (see, e.g., Tourangeau and Smith, 1996). Turner et al. (1998) provide a fuller review of CASI methods.

ACASI has been implemented for some modules on the British Crime Survey, but it has not been used in other victimization surveys, nor has it been used in other Census Bureau demographic surveys. However, it is notable that many federal government surveys contracted to the private sector that measure sensitive attributes use ACASI; these include the National Medical Care Expenditure Survey, the National Survey of Drug Use and Health, and the National Survey of Family Growth.

3–D CONSTITUENCIES AND USES: STATE STATISTICAL ANALYSIS CENTERS

Constituencies and consumers of NCVS data are varied and have diversified since the program's inception. Criminologists and federal justice policy researchers have historically relied on the NCVS to understand fundamental trend and victimization dynamics. However, in recent years the advent of a victim services infrastructure and increased public attention have widened the scope of interest in the NCVS specifically and victimization data more broadly. Contemporary users of the NCVS include:

- State justice statistics and services agencies;
- Victim services providers;
- Legislatures;
- State and local agencies, such as departments of health, mental health, and planning;
- Advocacy groups (e.g., domestic violence, child abuse, elder abuse, racial disparity); and
- The public.

Contemporary users are interested in victimization data that informs issues or problems of direct concern to them or their mission. This often means detailed findings on the incidence and nature of victimization in subpopulations, measures of change in the incidence and nature of victimization (trends), information on victimization in a specific geographic area (e.g., state, region, locality, neighborhood), and data on characteristics relevant to specific victims.

In this section—and in most of this report—we do not make as exhaustive a listing of constituencies and uses for the NCVS as our overall charge suggests. This is due to the initial interest of BJS in an examination of NCVS design options. In our remaining meetings and final report, the panel intends to canvass a fuller set of constituencies for BJS products. For this NCVS methodological report, we focus principally on use of victimization data by state-level statistical analysis centers and their related support organization, the Justice Research and Statistics Association.

3–D.1 SAC Network

Since its creation, BJS has supported state efforts to collect, analyze, and report criminal justice statistics through what was initially known as the Statistical Analysis Center (SAC) program. The SAC program was designed to foster criminal justice statistical infrastructure development in the states; its goal is to serve as a resource for policy formation and resource allocation by acting as a conduit for justice information between the federal and state governments and by providing additional information on the nature and dynamics of crime at the national and state levels. The SAC program did not provide resources to completely build state criminal justice statistical systems and clearinghouses but instead supported state efforts in this regard.

The program was redesigned in 1996 so that support to states would be for specific research or system development projects of mutual interest to states and the U.S. Department of Justice. The State Justice Statistics (SJS) program emerged to "maintain and enhance each state's capacity to address criminal justice issues through collection and analysis of data" (Bureau of Justice Statistics, 2006c:2).

A network of state SACs exists today in each state and two territories, created either by state statute or executive order. Although the size, location in government, and authorizing features of each SAC varies, they essentially serve similar functions and contribute to justice policy formation through research, support of legislative activity, executive policy development, and as a resource for state justice and related agencies. The SACs have also been important resources for BJS by providing assistance, data, and research at the state level on problems or issues of national interest. Individual state SACs also benefit from membership in the Justice Research and Statistics Association, the professional organization of state centers located in Washington, DC.

State SACs work closely with the criminal justice community and researchers in their state and are typically familiar with data systems, data quality, and information needs in their jurisdiction. The existing SAC network is familiar with the NCVS and the relevance of victimization research to the justice policy process. In some instances, SACs have conducted victimization surveys of varied methodological approaches in their own jurisdictions. A bibliography of recent reports related to victimization and victimization surveys solicited from the SACs is included in Appendix D, illustrating the ongoing interest in victimization at the state and local levels.

Our panel was informed in its work through a survey of SAC directors regarding the prevalence of victimization surveys conducted at the state or local level and the utility of the NCVS and victimization research for law and policy in their jurisdictions. In addition, three experienced SAC directors appeared before the panel to discuss the NCVS and victimization research needs.[3] Findings from the SAC survey indicate that victimization surveys are a valuable tool for policy makers and other users at the state and local levels. However, although the NCVS fulfills some of this need, it increasingly is not able to address issues of contemporary importance to victim services agencies, legislatures, advocacy groups, researchers, and governmental policy makers. Key findings from the survey are referred to below.

3–D.2   State Role for National-Level Data

NCVS data are used in a variety of ways by SACs and the agencies and organizations that work with them. Among the uses reported by SACs and evident in other published SAC reports are:

---

[3]Those appearing were Kim English, Colorado Division of Criminal Justice; Douglas Hoffman, Pennsylvania Crime Commission; and Phillip Stevenson, Arizona Criminal Justice Commission.

- State legislative support and testimony (the most common reported use);
- Benchmarking;
- Forecasting;
- Program and policy evaluation;
- Resource allocation;
- Victim services policy, planning and operations (particularly for initiatives funded under the federal Violence Against Children and Violence Against Women Acts);
- Victim experiences and satisfaction with the criminal justice system;
- Context and benchmarking in a broader study of recidivism patterns;
- Protocol development (e.g., the dynamics and relationship of victim-offender has been used to inform child abuse protocols in some states);
- Community oriented policing support and evaluation;
- Work with advocacy groups and other nongovernmental organizations (NGO); and
- Public and criminal justice system stakeholder education.

Although national-level estimates from the NCVS do not speak directly to rates and occurrences in local geographic areas, the state SACs still cite the utility of having some kind of national benchmark. The NCVS is used often for the policy and planning efforts of SACs and other constituents in the absence of state, regional, or local victimization data. Findings on trends and characteristics from the NCVS are often found in reports, briefs, and other documents for purposes of illustrating points important for policy and law formation and resource allocation.

Overall victimization rates remain of interest, but topical issues related to special victims have emerged as important for policy, planning, and service delivery in most jurisdictions. Of particular interest are domestic and sexual violence, factors related to the reporting of crime to police, and victim experiences with the criminal justice system.

3–D.3   Need for Finer Level Estimates

Although the survey of state SACs suggested continued interest in national estimates from the NCVS, it also clearly suggested a need and desire for victimization data and related information at the state and in some cases city or regional level. The national victimization measures contained in the current NCVS are generally useful as a triangulation tool, but they do not address the need for data at the state or local level given lack of the ability to disaggregate the findings. Users would find victimization data at the state, regional, and in some instances city or local levels more directly relevant to the policy and program uses encountered today.

State Victimization Surveys

In the absence of state-level estimates produced directly from the NCVS, about half of the state SACs have conducted a victimization survey of their own in recent years. Although the methods of these surveys vary somewhat, most replicate basic questions in the NCVS, primarily because those questions have been tested and validated over time and provide a basis for comparison with the NCVS. These subnational victimization surveys include efforts in Alaska (Giblin, 2003), Idaho (Stohr and Vazques, 2001), Illinois (Rennison, 2003; Hiselman et al., 2005), Kentucky (May et al., 2004), Maine (Rubin, 2007), Minnesota (Minnesota Justice Statistics Center, 2003), Pennsylvania (Young et al., 1997), South Carolina (McManus, 2002), Utah (Haddon and Christenson, 2005), and Vermont (Clements and Bellas, 2003). We have drawn our observations from the experiences reported in these states; additional information on some of these state efforts is given in Appendix D.

Basic observations from the state victimization surveys conducted to date include:

- Methods of data collection vary, but the surveys used either mail questionnaires (Idaho, Illinois, Minnesota, and Utah) or telephone interviews (Alaska, Kentucky, Maine, Oregon, Pennsylvania, and Vermont). Due to their one-time or semiregular frequency, none has attempted to replicate the NCVS panel structure of repeated interviews at the same addresses (or phone numbers) and have rather relied on cross-sectional samples.

- The number of respondents ranged from about 800 to 3,100. Response rates varied between 12 and 65 percent with no consistent pattern based on method of delivery.

- The state surveys generally do not attempt to estimate statewide or subpopulation rates from survey data. Most surveys report findings from within the sample.

- Surveys typically focused on general victimization experience to calculate the extent of victimization in the sample, often similar to the NCVS screener questions.

- All of the surveys were conducted on the adult population, primarily because of the legal and methodological difficulties associated with surveying younger groups.

- Most surveys collected data on special populations or issues of concern in the state (e.g. stalking, domestic violence, hate crimes, school victimization, disability, geographic area, gangs).

- Most surveys also examine perceptions of crime and public safety, fear of crime, and reporting of crime to police.

- Several surveys measured knowledge and use of victim services.

- Two states report using the BJS-developed Crime Victimization Survey software (described further below), effectively replicating the basic NCVS content.

Victimization surveys conducted at the state and local levels generally have not produced the level of statistical precision required for estimation used by the NCVS or similarly constructed surveys. Most have relied on sample sizes consistent with measuring public opinion, experienced mixed response rates, and generally measured self-reported victimizations without collecting detailed incident data. In most instances the studies have been conducted only once or for intervals in excess of one year, primarily for cost and administration reasons.

Our survey of SAC directors suggested that many would be greatly interested in conducting their own state or local victimization studies if it were practical to do so and if resources were available. However, mounting a survey is a costly proposition and is most often impractical for state agencies; it is especially impractical for local agencies and nongovernmental organizations. Most agencies, even at the state level, do not have the expertise required to design, implement, and analyze data from a statistically and methodologically valid survey. The ability to do so is further constrained by a lack of experience in conducting call center activities (for phone interviews), sampling design, and the availability of skilled analysts. As a consequence, replications of the NCVS at the state and local levels are not widely conducted. When studies have been carried out, it is often with the assistance of university-based researchers.

BJS has attempted to bridge this gap by developing desktop computer-based Crime Victimization Survey software. The software replicates many of the features of the NCVS survey and allows screening and detailed incident reports. Although it is a useful development, only a few states and localities have conducted their victimization surveys using the tool. Although the software product automates some parts of the process, mounting a state-level representative survey still requires personnel and resources that individual states have found difficult to obtain. The Alaska SAC report on the use of the BJS-developed tool for a victimization survey in Anchorage (Alaska Justice Statistical Analysis Center, 2002) reviews the basic features of the BJS-provided software and points out implementation problems raised during its early development.

Due to the resource demands, state victimization surveys tend to be one-shot or episodic events. However, a few states have conducted their own surveys on a more-or-less regular cycle (e.g., Minnesota's mail-based survey was conducted in 1992, 1996, 1999, and 2002). Consequently, the state surveys tend to be directed toward comparison with national NCVS trends; without a fuller time series of state estimates, they are limited in their ability to evaluate program and policy impacts at the state, regional, or local level.

It is important to note that the need for finer level data does not necessarily mean a strict disaggregation by state or other level of geography. Rather, state and local agencies like the state SACs would benefit from estimates based on samples that are "more like us"—demographically representative—in other respects than sheer geography. For instance, having more measures that can be disaggregated by level of urbanicity (urban, rural, suburban) would be useful and more relevant to individual jurisdictions than omnibus national totals. As an example, estimates from the Vermont Victimization Survey trended well with measures based on the NCVS sample from rural areas; hence, use of a "rural NCVS" analysis would be sufficient and more cost-effective than conducting an original study in Vermont on a regular basis.

Have Local Needs for Victimization Data Changed?

The demand for victimization data and research has significantly expanded since the NCVS was implemented in 1972. Perhaps the most important growth driver has been increased demand for more sophisticated and geographically disaggregated measures among state and local constituents. The contemporary rediscovery of crime victims and ensuing victimization movement parallels and is interwoven with the need for increasingly complex and textured victimization data (see Karmen, 2007). Aggregate national estimates of victimization rates and crime victim characteristics that were innovative at the time the NCVS was developed remain important for trend purposes, but they do not fully address needs that have emerged at the state and local levels.

In the decades following the development and implementation of the NCVS, the field of victimology and a victim services infrastructure have emerged, significantly fueled by federal, state, and private support. Karmen (2007:27–41), Walker (1998), and others have documented varied factors that in concert have contributed to expansion of victimology and victim services in recent years. Social visibility of vulnerable and politically underrepresented populations has propelled the need to understand victimization rates and patterns for various subpopulations as well as social stratification by race, class, and gender. Such forces as escalating crime rates in the 1960s and 1970s, the women's, civil rights, and children's rights movements, and elevation of domestic and sexual violence, and subsequent policy at the federal and state levels have pushed the demand for more data and research on crime victims. Such crimes as hate crimes or stalking, which were not part of the criminological lexicon when the NCVS was developed, illustrate how the environment and conceptualization of victimization have changed.

Understanding the general victimization rate for purposes of correlation with police-reported crime rates is still important at the state and local levels, primarily for assessing crime trends and patterns. However, more detailed and segmented information about victimization patterns is often needed to craft policy, services, and resource allocation. Contemporary victimization issues include understanding victimization across different population segments, some of which are vulnerable and of significant public concern and have been addressed in the NCVS through topic supplement surveys, like the School Crime Supplement and the Police-Public Contact Survey supplement (see Demographic Surveys Division, U.S. Census Bureau, 2007a).

The NCVS topic supplements have provided significant new data and information and as such have been important innovations to the NCVS. However, they are also—as currently implemented—adjuncts or add-ons to the main NCVS and hence may not necessarily reflect the type of sample that would ideally be drawn to study the subject. The need for supplemental surveys reflects contemporary demand for enhanced victimization knowledge and should be reexamined relative to the continued role and form of the NCVS.

Significant state and federal resources have helped shaped the victims movement over the past three decades and consequently have indirectly fueled the need for more rich and geographically focused victimization data. Federal resources have been provided for states directly through landmark legislation such as the Victims of Crime Act of 1984 (VOCA; P.L. 98-473 §1401 et seq.) and the Violence Against Women Act of 1994 (VAWA, reauthorized in 2000; P.L. 103-322 and 106-386). The VOCA legislation created the Office for Victims of Crime in the Office of Justice Programs, U.S. Department of Justice, and has assisted states in various ways to construct victim services and compensation fund infrastructures. The VAWA continued efforts in this area by providing resources to improve the investigation, prosecution, processing, and restitution enforcement for victims of crime. The National Center for Victims of Crime has also emerged as a central nongovernmental resource in this movement since 1985 (National Center for Victims of Crime, 2003).

The Office for Victims of Crime has grown into a significant resource and facilitated development of a victim services and enforcement infrastructure at the state level. The demand for victimization data, information, and research has grown exponentially as programs develop, service resources are allocated, and programs and policies are evaluated. Most federal justice grants have required evaluation components for at least a decade, another source of increased demand for victimization research at the state and local levels.

One side note related to efforts to understand victim characteristics and victimization is in order. Some states and jurisdictions have implemented incident-based reporting systems and, specifically, systems compliant with the National Incident-Based Reporting System (see McManus, 2002). Contemporary records management and incident-based systems hold promise for capturing victim data linked to offenders and crime characteristics, but this promise is yet to be realized on any scale. While NIBRS and similar data may comment on victim-offender relationships and characteristics, these data remain constrained, since they represent reported offenses. In many jurisdictions, however, incident-based or NIBRS data are the only small area victim data available and have been used for policy, planning, and evaluation in the absence of comprehensive victimization data.

Legislative and Executive Support Crime policy bills are quite prevalent in legislatures around the country, with few actually making it into law. However, extensive debate occurs and requires reasoned analysis. Policy is often driven by celebrated cases, and crime data are needed to debunk myths or unusual circumstances. The dangers of the lack of information are less effective policies and poor allocation of state and federal resources.

Having data over time is also extremely important, although the desirable time intervals of measures may vary. In some rural jurisdictions, victimization patterns may not change quickly enough to warrant annual surveys. Victimization data may not be able to comment specifically on the efficacy of particular programs, but over the long term these data are critical to understanding larger impacts of policies on crime and victimization patterns.

3–E VALUING VICTIMIZATION INFORMATION: COMPARING THE COST OF VICTIMIZATION MEASUREMENT WITH BENCHMARKS

Arguably, the most significant challenge faced by the NCVS—and largest constraint on its survival—is the availability of funding resources. As described in Chapter 1, BJS has been subject to essentially flat funding for a number of years, constraining options on the NCVS as the cost of conducting the survey has grown. Accordingly, it is important to consider the question of the value of the information that the NCVS provides. This can be done formally, as suggested by the framework outlined in Box 3-1. In

---

Box 3-1   Value of Information from a Decision-Making Point of View

A useful perspective in designing or redesigning a public statistical system or survey is to consider the value of information. The perspective is valuable because the purpose of the system is to improve the efficiency of actions by the government and other users of the data. Empirical valuation of information is extremely difficult (National Research Council, 1976a; Savage, 1985) for a variety of reasons: uses may not be identified, uses may be identified but the role of data may be imperfectly understood, valuation of alternative choices under different states of nature may be infeasible. Some examples in which the valuation of information may be feasible are discussed by Spencer (1982).

The basic idea for the value of a survey such as the NCVS may be illustrated by the following stylized example. Suppose that in the absence of NCVS data, alternative actions A_1, A_2,...,A_m would be taken with respective probabilities p_1, p_2,...,p_m. With the NCVS, the alternative actions A_1, A_2,...,A_m are taken with respective probabilities q_1, q_2,...,q_m. The expected value of information for this use alone may be represented as

$$(p_1 - q_1)U(A_1) + (p_2 - q_2)U(A_2) + ... + (p_m - q_m)U(A_m)$$

where $U(A_1)$ is the expected value if action A_1 is taken. The information is valuable if it leads to higher probabilities of more valuable actions being taken. Differences in values of alternative actions may reflect the differences in value of passing one law (or one version of a law) rather than another. Even if dollar valuation is not feasible, a sense of the impact of alternative laws may lead to a sense that the difference in value is on the order of tens or hundreds of thousands of dollars, or perhaps more.

---

this section, we take a practical approach to assessing the value of NCVS information by comparing the cost of the NCVS with several relevant benchmarks: estimates of the fiscal cost of crime, the costs of other federal survey data collections, and the expenditures of other countries in measuring victimization.

3–E.1   The Cost of Crime

The total cost of crime in the United States—including both tangible economic costs and intangible costs and covering such components as damages to victims and expenditures on the justice and correctional systems—is an elusive quantity to estimate. A large research literature has tried to estimate the economic costs of crime, and we briefly summarize some points from this work in this section. There is certainly a large speculative element to these figures, and the calculation of intangible costs is especially uncertain. In raising the cost of crime as a comparison benchmark for the NCVS, we do not suggest that the costs of crime and the costs of victimization measurement should be directly linked (e.g., that spending on the NCVS should be some set fraction of the cost of crime). Instead, we offer the comparison for two purposes. The first is to reinforce the idea that crime is a sufficiently important and complex phenomenon facing the United States as to warrant multiple, complementary, and detailed statistical indicators (i.e., both the NCVS and the UCR, as discussed further in Section 3–F). The second is to highlight a unique and important substantive function of the NCVS: the survey is the only direct, systematic source of information on victims' economic losses due to crime.

Studies of the economic cost of crime often arise in the context of a benefit-cost analysis in which criminal justice spending is weighed against the economic losses associated with victimization. Gray (1979) provides a historical review of research on the costs of crime and traces the earliest studies to the early twentieth century. Cohen (2000, 2005) provides a comprehensive literature review and analysis.[4] This section draws heavily from the discussion by Cohen (2000).

Research on the costs of crime distinguishes at least nine different types of costs: (1) direct property losses; (2) medical and mental health care; (3) victim services; (4) lost workdays, school days, or days of domestic work; (5) pain and suffering; (6) loss of affection and family enjoyment; (7) death; (8) legal costs associated with tort claims; and (9) long-term costs of victimization. Some of these costs accrue directly to crime victims and their families. For example, the cost of lost property that is unreimbursed by insurance is borne by the victim. Other costs are socially distributed. For example, losses reimbursed by insurance are passed on to society in the form of higher premiums.

These costs can be categorized broadly as either tangible or intangible. Tangible costs involved monetary payments, such as medical costs, stolen or damaged property or wage losses. Intangible costs are nonmonetary and include things that are generally not priced in the marketplace, like pain and suffering or quality of life. In principle, tangible costs are relatively straightforward to estimate, but great uncertainty accompanies the estimation of intangible costs.

Although the calculation of tangible costs is conceptually straightforward, Cohen (2000:282) reports that the NCVS provides "the only direct source of crime victim costs." The NCVS obtains from crime victims dollar estimates of the costs of medical care, lost wages, and property loss (Klaus, 1994). These figures are likely to understate the total tangible cost because the recency of the victimization reference period excludes longer term medical costs. In addition, the survey does not count mental health costs or other less proximate costs, like moving from the neighborhood or buying home security systems. Some estimates indicate that the tangible costs of victimization are higher than those recorded by the NCVS by a factor of 4

---

[4]Anderson (1999) also reviews previous studies of the cost of crime. Attempting to estimate indirect and opportunity costs associated with crimes, Anderson suggests that the annual net cost of crime in the United States is about $1.1 trillion.

Table 3-2 Estimates of the Average Economic Loss
Associated with Criminal Victimization

		Miller et al. (1996)		
Crime	Klaus (1994)	Tangible	Intangible	Total
Rape	$234	$4,962	$79,202	$84,164
Robbery	555	2,238	5,546	7,784
Assault	124	1,508	7,589	9,097
Theft	221	360	0	360
Burglary	834	1,070	292	1,362
Auto theft	3,990	3,406	292	3,698

NOTE: Intangible costs are estimates of lost quality of life. All
figures are in 1992 dollars.

for robbery, a factor of 10 for assault, and a factor of 20 for rape (Miller
et al., 1996). Other tangible costs of crime are missed entirely by the NCVS.
White-collar crimes like fraud or theft of services are difficult to quantify
because victims may not be aware of the crime. Potential victims also suffer
(unmeasured) tangible costs in the form of crime prevention expenditures.

Intangible costs of pain and suffering and lost quality of life are even
more difficult to estimate. Some studies have tried to capture the intangible
costs of crime by studying the relationship between index crime rates and
housing prices. These studies see the risk of victimization as capitalized in
housing prices (Thaler, 1978). In another approach, Cohen (1988) used jury
awards in tort cases to estimate the monetary value of pain and suffering
and lost quality of life. Zimring and Hawkins (1995) criticized this and
related work for its arbitrary measurement of intangible costs. Alternative
to jury awards, such as workers compensation payments, might have been
used, yielding alternative estimates of the intangible costs of crime. In short,
intangible costs are highly uncertain and difficult to quantify.

Table 3-2 reports a range of estimates of the dollar cost of crime. The
table compares the economic losses reflected in the NCVS reported by Klaus
(1994), with those calculated by Miller et al. (1996), which include a more
expansive inventory of costs. Klaus (1994) uses just those tangible losses
reported in the 1992 NCVS. Miller et al. (1996) partly base their estimates
of tangible costs on the NCVS, although they add estimates of mental health
care and lifetime medical costs, as well as long-term productivity losses.
Intangible costs are based on adjusted jury awards for pain and suffering.
Clearly, estimates based on a broader consideration of costs yield far higher
estimates than the NCVS alone. For violent crimes, intangible costs domi-
nate estimates of the total economic loss.

The large average cost of tangible and intangible losses sum to large losses in the aggregate. In the aggregate, Klaus estimates that crime victims lost a total of $17.6 billion in direct costs. Miller et al. (1996) report that the economic cost of index crimes in 1990 summed to $450 billion, in 1992 dollars. Of this total, $345 billion was due to lost quality of life, and $105 billion was due to tangible economic losses. Fatal crimes, including drunk driving incidents and rape together account for $220 billion.

Using almost any of the above metrics, criminal victimization is one of the key attributes affecting the progress and status of a modern society. It is fitting, therefore, that the authorizing legislation of BJS gives to it the mandate to measure victimization, as a key social indicator of the country's progress.

3–E.2   Comparison with Other Federal Surveys

The National Crime Victimization Survey is conducted largely from the 12 regional offices of the U.S. Census Bureau. The Census Bureau also conducts the labor force survey, the Current Population Survey, for the Bureau of Labor Statistics. It conducts the National Health Interview Survey for the National Center for Health Statistics. It conducts the American Housing Survey for the Department of Housing and Urban Development.

It also conducts periodic surveys, for example, the National Household Travel Survey for the Bureau of Transportation Statistics and the American Time Use Survey for the Bureau of Labor Statistics. These often use the Current Population Survey as a convenient sample for reinterviewing on special topics.

The data collection costs for these surveys are sometimes difficult to discern, although rough estimates can be constructed from the presentation made by Census Bureau staff to the panel at our April 2007 meeting. The cost per interview for the NCVS in fiscal year 2006 was estimated at $146;[5] at the 2005 rate of 38,600 households interviewed, this would imply total costs of $5,635,600.

In the judgment of the panel, the appropriate criterion for assessing how much the country should spend on victimization measurement is the fitness of NCVS estimates for their uses. Fitness for use criteria would entail the BJS articulating all uses and placing them in the context of importance of the uses for the country. These are inherently value-laden judgments. BJS needs some benchmarks for such judgments. They might be had by comparisons with other federal statistical agencies data series.

---

[5] By comparison, the cost per case for the National Health Interview Survey in fiscal 2006 was estimated as $212 by Census Bureau staff; the cost of a Current Population Survey interview as $64.

3–E.3   International Expenditures

Comparisons of what the United States spends on crime statistics and what other similar nations spend provides an alternative standard for assessing the sufficiency of U.S. expenditures in this area. With this said, making cross-national comparisons is not simple. One must find nations with similar resources and infrastructure and with similar expectations about public safety and governmental accountability. Even when these larger institutional structures are similar, arcane budgeting procedures can complicate comparing expenditures. Nonetheless, if one can negotiate these rapids, cross-national comparisons can be very illuminating.

In terms of identifying nations with basic social and political institutions similar to the United States, it would seem that most western, industrialized democracies would be fitting comparison points. Nations in Western Europe, Australia, and Canada are a good set of comparison points.

In addition to simply comparing budgets for collecting crime and justice statistics across these nations, it may be useful to standardize these expenditures by some features of these nations that could reasonably be assumed to affect the cost of collecting and reporting these data. Population size, for example, may increase the cost of collecting and reporting crime statistics. Larger nations have more correctional facilities, so any census of these facilities would include more facilities and more funds. There are ways to reduce these costs, but, in general, it is not unreasonable to assume that the larger the population, the greater the cost of crime statistics. Similarly, the land mass of a nation can affect the cost of collecting crime statistics. To the extent that in-person visits are required, data collection in far-flung places will entail more travel costs or the maintenance of a standing field staff that would not be required in smaller places. The volume of crime will also influence the collection of data on crime. A nation with 10 crimes should have fewer transactions to document than a nation with 100,000 crimes. So standardizing crime statistic budgets by residential population, land mass, and the volume of crime will make for more comparable data across nations.

There are a number of other differences between nations that are clearly relevant for cross-national comparison, such as the degree of administrative centralization in a country or the nature and extent of federalism. These differences are perhaps more consequential than the ones noted above, but we are not yet in the position to standardize comparisons for these effects.

At this time we make comparisons only between the United States and England and Wales; these expenditures are shown in Table 3-3. Moreover, we have restricted comparisons to the costs of collecting victimization survey data because the collection of court and corrections data in England and Wales now resides with the Ministry of Justice and not the Home Office.

Table 3-3 Comparative Expenditures on Victimization Surveys, United States and England and Wales

Nation	Total Expenditures on Victimization Surveys	By Population per 1,000	By Square Kilometer	By Number of Serious Crimes
England and Wales	12,500,000.00	212.62	82.70	13,897.41
United States	20,731,800.00	73.67	2.26	4,336.91
Ratio of E&W to US	0.60	2.89	36.55	3.20

SOURCE: Land area and population data derived from
http://www.nationsencyclopedia.com/economies/Europe/United-Kingdom.html and
https://www.cia.gov/library/publications/the-world-factbook/print/us.html.

In fiscal year 2006, BJS spent $20.7 million collecting, processing, and reporting NCVS data. The Home Office spends approximately $12.5 million doing the same for the BCS. The United States has roughly four times the population of England and Wales; so on a per capita basis, the former spends $73.67 per 1,000 population on victimization data while the latter spends $212.62. England and Wales spend almost three times as much as the United States. When viewed in terms of land mass, the differences are even greater. The mainland United States is 9,161,000 square kilometers and England and Wales are 151,000 square kilometers. On a per square kilometer basis, England spends almost 36 times as much on victimization statistics as the United States. If we examine these expenditures by police-recorded serious crime volume, England and Wales spend more than three times what the United States spends on victimization statistics. This difference is about 10 percent greater than what we observed by population alone. These comparisons suggest that—at least compared with one international benchmark—the collection of victimization statistics in the United States has been given relatively less funding compared with England and Wales.

In making a comparison with the experience of England and Wales, it is worth noting that a particular role has been defined by statute for the BCS; this formalizes a use and a constituency for it—and adds justification for expenditure on the survey—in a way that does not exist for the NCVS. The Local Government Act 1999 created a set of indicators that are used to measure the performance of government departments and local authorities; the indicators are periodically revised. These indicators are formally known as "best value performance indicators"; in the area of policing, they are commonly described as "statutory performance indicators" or SPIs.[6] The SPI data are collected and audited annually by the Audit Commission and are also published on government websites. BCS data are formally required for several of these indicators (Home Office, 2007): for example, in the set of SPIs defined for 2006–2007, "the percentage of people who think their local police do a good job" (SPI 2a), "perceptions of anti-social behaviour" (SPI 10b), and the violent crime rate (SPI 5b). Meeting these statutory guidelines requires that the BCS be regularly funded and capable of providing estimates at the local government level.

3–F   ISSUES RELATED TO THE COEXISTENCE OF THE NCVS AND THE UCR

For more than three decades, the nation has had two national indicators of crime: the Uniform Crime Reporting program and the National Crime Victimization Survey. As described in Chapter 2, the two programs overlap in the crimes they cover (and both are used to generate national-level estimates of violent crime) but also differ in some important definitional ways. Despite the definitional differences between the two measures and their complementary nature, a fundamental question still arises in public discussions of crime statistics: Is it necessary to have two data systems for the purpose of estimating and evaluating trends in crime?

One part of that broader underlying question concerns the trends shown by the two series and the degree to which they agree or converge over time: In other words, do police records generally reflect victimization trends, and vice versa? A second part of the bigger question is more philosophical, concerning the necessity of two series: Does there remain the need for a second indicator completely independent of the official police reports?

3–F.1   Do Police Record Reports Reflect Victimization Trends?

The question about the concurrence of the NCVS and UCR trends is made more salient by the fact that there appears to have been a convergence in recent years of UCR- and NCVS-based national estimates of serious violent crime (i.e., rape, robbery, and aggravated assault). In other words, estimates from the NCVS of the number of crimes victims say they have reported to the police and the number that are recorded in the UCR program have grown closer in recent years (see Figure 3-2). On its face, the evidence of the most recent years of the series might suggest a redundancy—that national crime trends may be adequately described by UCR and that the NCVS role as a crime trend monitor may have diminished.

---

[6]Additional information on best value performance indicators is available at http://www.bfpi.gov.uk/pages/faq.asp.

The correspondence or divergence of the UCR and the NCVS takes on extra importance when there appears to be a shift in long-term crime trends. Between 1992 and 2004 the United States experienced a long period of declining crime, by both measures. More recently there have been signs of an upturn in crime in a number of cities, and the national crime rate as measured by the UCR has increased in two consecutive years. It is precisely at turning points such as the current one that rich and timely measures of a variety of aspects of crime are required, to better understand the trajectory that the nation seems to be taking.

Although the UCR and NCVS estimates of the number of serious violent crimes reported to the police have generally become more similar over the past decade, this convergence in levels is not yet fully understood, nor is it clear whether the similarities in estimates will continue in the future (see Lynch and Addington, 2007). This is because the convergence in the estimates does not necessarily reflect a reduction in error by one or both series. Rather, the two series can produce different estimates and varying long- and short-term trends because they measure different aspects of the crime problem using dissimilar procedures. Furthermore, even if the convergence does reflect some reduction in error associated with one or both series, there is little evidence to suggest that this pattern will remain constant in the future.

As evident in Figure 3-2, annual estimates of the total number of serious violent crimes derived from NCVS data have often been higher than the annual counts in the UCR. There are several reasons why this may occur. Most importantly, the NCVS data include crimes that are not reported to the police. Approximately 49 percent of violent victimizations and 36 percent of property victimizations are reported to the police (Hart and Rennison, 2003). In addition, NCVS counts may be higher if police departments do not record all of the incidents that come to their attention or do not forward the reports to the national UCR program.

For some types of crimes in the NCVS and the UCR, it is possible to reconcile apparent discrepancies in annual estimates by adjusting the NCVS counts to include only those incidents said to have been reported to the police. When such adjustments are made, levels and trends in burglary, robbery, and motor vehicle theft appear generally similar in the NCVS and UCR. However, UCR and NCVS levels and trends in serious violent crime, such as aggravated assault and rape, exhibit many discrepancies after these kinds of adjustments are made. These differences in both levels and trends in aggravated assault and rape may result from changes concerning the public's willingness to report crime to the police, changes in the way police departments record crime, or some other factor. It is clear that the differences in the methodologies of the UCR and NCVS programs must be considered when assessing both levels and trends of crime in the nation. However, the fact that the extent of agreement in current levels of crimes depends on the

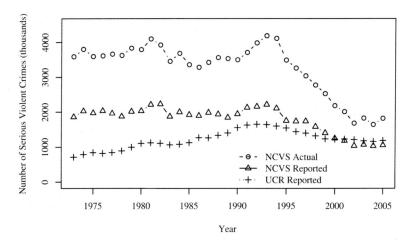

Figure 3-2 National Crime Victimization Survey and Uniform Crime Reports estimates of serious violent crimes, 1973–2005

NOTES: Serious violent crimes include rape, robbery, aggravated assault, and homicide; homicide estimates from the UCR are added to the NCVS series. "NCVS Actual" includes crimes not reported to the police as well as those that are ("NCVS Reported"). NCVS estimates before 1993 are based on data year; for 1993 and later years collection year is used (see Table C-2).

SOURCE: National Crime Victimization Survey, Bureau of Justice Statistics and Uniform Crime Reports, Federal Bureau of Investigation. Data from http://www.ojp.gov/bjs/glance/tables/4meastab.htm [11/1/07].

nature of the offense makes it difficult to claim that UCR data alone could be a sufficient indicator for estimating current levels of "crime."

If the goal is to assess long-term trends in crime, then a high correlation between UCR and NCVS trends would suggest that either data series would serve as a reasonable proxy for some analytical purposes. McDowall and Loftin (2007) assessed the correlations between UCR and NCVS national trends for index crimes for the period 1973–2003. Using a correlation standard of 0.80 or higher to indicate sufficient agreement in trends, they found that only two crimes came close to or exceeded this standard: robbery ($r = 0.76$) and burglary ($r = 0.93$). The next highest correlation was found for motor vehicle theft ($r = 0.67$). However, the remaining crime types exhibited much lower or even negative correlations. For larceny theft, the correlation was weak ($r = 0.20$), and for rape and assault, the correlations were negative ($r = -0.16$ and $r = -0.21$, respectively) (McDowall and Loftin, 2007:101). Like current level estimates, the trend correlations varied according to crime type. Analysts studying robbery and burglary can expect generally similar results using either UCR or NCVS trend data for this time period. For other types of crime, however, this will not be the case.

One of the main hypotheses about why assault trends might differ in the UCR and NCVS series is heightened police productivity resulting in a growth of police estimates of assaults (O'Brien, 1996). Rosenfeld (2007) hypothesized that if the divergence in the two series was the result of changes in the way police were handling less serious assaults, then one should expect to see similar trends in the UCR and NCVS gun assault rates, but divergent trends in the nongun assault rates because the perceived seriousness of gun assaults and the ways in which such crimes are handled by the police are much less susceptible to change over time. In addition, if the hypothesis is correct, the ratio of nongun to gun assaults should have increased more in the UCR than in the NCVS. Using UCR and NCVS aggravated assault data for the period 1980–2001, Rosenfeld found that the correlation between the UCR and NCVS estimates of gun assaults was 0.74, while the correlation for nongun assaults was 0.16 (not significantly different from 0). In addition, the ratio of nongun to gun assault rates in the UCR grew, while the same ratio using the NCVS data did not. Thus, the two data series provided similar information about trends in gun-related aggravated assault, but they differed in their patterns for aggravated assaults without guns. The form of the UCR and NCVS nongun assault trends also suggested that changes in police recording and categorization of such incidents stabilized during the 1990s as the two series began to exhibit more similar trends.

In their comprehensive examination of UCR and NCVS trends, McDowall and Loftin (2007) found that the two series for each of the index crimes began tracking each other more closely in the 1990s. McDowall and Loftin argue that this would suggest a structural break in the UCR data series, which might indicate that the estimates from the two series will continue to follow each other more closely in the future. The reason that modifications in the UCR are thought to be responsible for the increased correspondence is that there have been changes in the recordkeeping systems of police departments, while the NCVS methodology remained relatively more stable over the same period (McDowall and Loftin, 2007:111). The recordkeeping capacities of police departments improved as a result of technological innovations, as well as increased numbers of personnel involved in this task. However, the authors note that the agreement in the series is fairly recent and based on a limited number of data points: thus it is premature to conclude that it will continue in the future (McDowall and Loftin, 2007:114).

Whether it is reasonable for state and local governments to believe that their local police data accurately capture trends in crime is more contentious. When state and local governments are interested in assessing trends in crime in their own areas, they typically must rely solely on police data because victimization survey data are rarely available for small geographical areas. The collection of reliable crime survey data is costly, and most state and local governments have not had the resources to conduct their own victimization surveys, especially on an annual basis. As a result, many wonder whether conclusions about the recent convergence in national police and victim survey data apply to their local areas.

The limited amount of research that has addressed the comparability of UCR and NCVS trends in local areas has used data from special tabulations of NCVS data. One such special subset allows researchers to produce victimization estimates for the 40 largest metropolitan core-county areas in the country (Bureau of Justice Statistics, 2007b). NCVS estimates can be generated from this newly available file for comparisons to UCR data for those same places and years (Lauritsen and Schaum, 2005). Existing research using these data has found that the correlations in the trends for robbery and aggravated assault vary considerably across metropolitan areas, although much less so for robbery than for aggravated assault (Lauritsen, 2006a). For robbery, the average UCR-NCVS trend correlation across the 40 largest metropolitan areas for 1979–2004 is 0.59 (range = 0.02 to 0.95), while the average correlation for aggravated assault is 0.16 (range = −0.75 to 0.76). In addition, the trend correlations tend to be higher in the more populated metropolitan areas, which may reflect earlier adoption of crime records management technology by the larger police departments. Lauritsen concluded that there is a good deal of variation at the level of the metropolitan statistical area in the correlations between the two sets of trends that is masked at the national level, and, as a result, it would be unwise for local areas to assume that their local UCR data provide good indicators of nonlethal violence trends. In addition, Wiersema (1999) developed an area-identified NCVS data set from public-use data files, coded with geographic identifiers down to the census tract level, that was briefly available through Census Bureau research data centers. These area-identified data have been used to support some subnational analyses; see, e.g., Baumer (2002); Baumer et al. (2003); Lauritsen (2001). However, the data have been taken out of circulation; we discuss this further in Section 5-A.

In sum, although much less is known about how UCR data trends compare with NCVS trends for state and local areas, it appears that at the national level, index crime trends have become more similar in recent years. However, just as structural breaks in UCR data collection appear to be responsible for the increasingly similar trends in the 1990s, changes may occur in the future. These alterations can result from a variety of factors, ranging from resource shortages in police departments to local political pressures regarding crime rates. Changes in the NCVS estimates can occur as well, as a result of declines in participation rates, sample coverage problems, limited resources, or other factors. Without both sources of crime information, it extremely difficult to fully understand the meaning of future changes in crime rates from either data series.

McDowall and Loftin (2007) and others (e.g., Lynch and Addington, 2007) argue that too much emphasis should not be placed on the general issue of convergence in national crime trends. Rather, the complementarities of the two data systems should be emphasized and the selection of one series over another should depend on the research question at hand. For some types of analyses, researchers can use both data sources to assess and understand the strengths and limitations of findings. However, the NCVS is currently the only available source of national data for describing and understanding trends in certain types of crime. These crime types include those that are defined according to specific conditions of the incident (such as intimate partner violence); victim characteristics (such as violence against women or crimes against the elderly); and crimes that are believed to be severely underrepresented in police data for assorted reasons (such as hate crimes, sexual violence, and identity theft).

3–F.2 Independence of the NCVS and the UCR

The origin of the NCVS as an *independent* estimate compared with the UCR was a development from the social and political climate of the 1960s. Cantor and Lynch (2000:97) note that "the confluence of several forces"—including a general mistrust of institutions—"made the 1960s an auspicious time for the development of victim surveys." Specifically, "reforms of several of the Nation's metropolitan police departments were accompanied by exposés of the previous practice of killing crime on the books"—that is, suppressing levels of reported crime. Against this backdrop, "victim surveys brought the 'patina of science'" and an air of accuracy and impartiality to crime statistics; "there was greater trust that the resulting [NCVS] crime estimates were not purposely manipulated" because "the Census Bureau and survey research agencies were not interested parties with respect to the crime problem" (Cantor and Lynch, 2000:88–89).

Beyond the question of whether the UCR and the NCVS respond to the same underlying phenomena, the question can be raised about whether the need for a independent, national-level, and victimization-based measure of the traditional index crimes persists. If it were concluded that there was no need for such an independent national-level measure, then a different class of NCVS design options becomes feasible, if not preferable: for instance, crime-type coverage between the NCVS and the UCR would be reallocated so that the UCR becomes the sole source of national indicators of some crimes, while the NCVS is focused more on hard-to-measure or newly emerging crime types.

We think that the NCVS has strong policy relevance as a national-level measure of crime independent of the UCR and should continue to function as such, for several key reasons. These include:

- *The NCVS as an objective measure:* Given that police are not an uninterested party in crime rates, there is always an inherent possibility of minimizing or reducing reported crime. Put more colloquially, having the police as the reporter of crime suggests the possibility of "cooking the books," lowering reported counts or possibly declining to report altogether. The UCR can do some imputation (and does), but an independent measure as a counter to this possibility still has merit. (Note, though, that this argument is weakened in the absence of local-area NCVS-based estimates, which would be the best check on individual department reporting.)

- *Voluntary UCR reporting leads to coverage gaps:* The UCR program relies on the voluntary cooperation of more than 17,000 seprate law enforcement agencies. Complete nonresponse to the UCR program, for individual years or for long stretches of time, occurs and is sometimes pervasive for some states and large localities (see, e.g., discussions of UCR coverage in Maltz, 1999, 2007). Again, imputation helps to bridge gaps in national-level UCR estimates, but for representativeness in coverage, the conceptual advantage still goes to a nationally representative sample like that employed in the NCVS.

- *Independent measure as "calibration" device:* It is useful to have two related-but-not-identical measures in simultaneous operation simply because they may not always agree. The United States has two independent measures of jobs and employment (in the Current Employment Statistics and the Current Population Survey); it has multiple measures of health insurance prevalence and of disability. The differences among the indicators enhance general understanding of the dynamics of the phenomena under study. Divergent or discrepant findings resulting from the two series may signal some structural problem with either of the individual measures and draw attention to potential problems in methodology. An original implicit notion in the creation of the NCVS—grounded in distrust of police reporting—may have been the use of the NCVS as a check on the UCR; however, it is equally valid to say that the two series can serve as an operational and conceptual check on each other.

At the same time, in speaking of "calibration," it is important to bear in mind that one data source is not always unequivocally right and the other wrong; both the NCVS and UCR are subject to measurement flaws. UCR measurement can suffer from lack of reporting by law enforcement agencies (discussed below) and underreporting due to victims' hesitance to come forward to the police. But, likewise, crimes may not be reported to interviewers either, as discussed in Section 3–B.2; other examples of crimes missed in NCVS exist, including the finding by Cook (1985) that National Crime Survey estimates captured only about one-third of gun assaults resulting in gunshot injury that were apparent in emergency room data. Such underestimates might arise from disproportionate gunshot prevalence among those not part of the household population or not listed as household members (and thus not sampled), those who were never contacted or who refused to participate, as well as those respondents who failed to report incidents to the NCVS interviewer.

Having an independent measure is important as long as there remains reason to believe that not all crime is reported to police and that not all crimes known to the police are completely tallied in the UCR. That said, the utility of an UCR-independent measure of crime should not prevent consideration of design options that reduce lockstep similarity between the UCR and the NCVS (e.g., measuring exactly the same set of "index crimes" except for homicide).

For several years, the National Incident-Based Reporting System (NIBRS) has been developed as a next-generation version (and replacement for) the UCR. The presence of a strong and complete NIBRS program might further blur the line between the UCR and the NCVS as separate indicators of crime. However, NIBRS development has been slow, and its coverage (i.e., cooperation by agencies in providing more detailed incident reporting) is still quite small. As of September 2007—about 15 years after development of initial NIBRS protocols—only about 26 percent of law enforcement agencies that contribute data to UCR were submitting NIBRS-compliant information; see Section D–2 for additional detail.

3–G ASSESSMENT

As is true of many multipurpose social indicators, the basic utility of the NCVS to the American public is difficult to characterize in tangible terms. Because it does not currently provide estimates at small areas of geography, its role in allocating federal or state funds for criminal justice improvements is limited, and it does not readily lend itself to focusing specific police interventions in specific neighborhoods. However, through its focus on providing detail on all types of crime and violence—reported to the police or not—and its rigorous design based on a representative sample with uniform national coverage, the NCVS has undeniable importance as a critical statistical indicator. For an informed assessment of the state of public welfare, federal statistical agencies like BJS have a core mandate "to be a credible source of relevant, accurate, and timely statistics" (National Research Council, 2004:3). The NCVS provides information on the extent, consequences, and causes of violent behavior that are not available at the same level of comprehensiveness and quality from any other source. Accordingly, direct reports from BJS on NCVS trends are frequently sought for information and for assessment of new policy, and NCVS data play an important role in national appraisals of child welfare (Federal Interagency Forum on Child and Family Statistics, 2007) and public health (U.S. Department of Health and Human Services, 2000).

In our assessment, the need for a victimization-based measure of crime—another indicator, separate from the official police reports of the UCR—is as significant today as it was when the NCVS was first conceptualized. This is the case not out of any inherent distrust of official reports to police or demonstrated inaccuracy therein, but rather for the reason suggested most concisely by the President's Commission on Law Enforcement and Administration of Justice (1967:18): "No one way of describing crime describes it well enough." The importance of the crime problem in the United States demands ongoing monitoring from multiple perspectives. In this monitoring, the NCVS is a vital complement to the police reports of the UCR, providing valuable information on the context and causes of victimization in ways that summary counts can never do by themselves (and in which even a fully implemented NIBRS would still be lacking).

However, in its size and available resources, the current NCVS is not capable of matching the original vision of the survey. As the costs of collecting information from the U.S. public have risen, the NCVS budget has not kept pace. Budget reductions have led to cutbacks in NCVS activities, most often through cuts in the total sample size. As it is currently configured, the NCVS does not meet the goal of being able to accurately measure year-to-year change in crime trends. That is, the standard errors of change estimates are too large to detect changes of importance to the country; BJS has had to use averages from 2-year groups of data in order to make statements about change (see Catalano, 2006), even though inferences from these rolling averages are not as intuitive to users and members of the public as direct estimates of change.[7] To state this as a finding:

> Finding 3.1: As currently configured and funded, the NCVS is not achieving and cannot achieve BJS's legislatively mandated goal to "collect and analyze data that will serve as a continuous and comparable national social indication of the prevalence, in-

---

[7]For additional information on the use and interpretation of rolling-average estimates from federal survey data, see National Research Council (2007); the American Community Survey will use 3-year and 5-year averages in order to produce estimates for small areas and populations.

cidence, rates, extent, distribution, and attributes of crime . . ."
(42 U.S.C. 3732(c)(3)).

Clearly, given the panel's charge to consider options for the conduct of
the NCVS, one possibility is *not* to conduct the NCVS at all; we reject that
option. To take that option violates the legislative responsibilities of the
Bureau of Justice Statistics. Furthermore, the panel thinks that BJS is the
appropriate locus of responsibility for victimization measurement. As a federal statistical agency, it alone has the mandate for independent, objective,
statistical measurement, with the transparency that can establish public trust
in the information (see National Research Council, 2004).

Thus, although there should be no need for the panel to do so, we feel
obliged to state a recommendation that is already explicit in the mission of
BJS:

> *Recommendation 3.1:* **BJS must ensure that the nation has qual-
> ity annual estimates of levels and changes in criminal victim-
> ization.**

However, the natural corollary is that the resources necessary to adequately achieve this mission must be forthcoming:

> *Recommendation 3.2:* **Congress and the administration should
> ensure that BJS has a budget that is adequate to field a survey
> that satisfies the goal in Recommendation 3.1.**

NCVS' unique substantive niche is providing information on crimes that
are particularly likely to go unreported to the police, for whatever reason—
whether fear or stigma, individual distrust of authority, or the perception
that a violent act or threat is not significant enough a "crime" to report.

> *Recommendation 3.3:* **BJS should continue to use the NCVS to
> assess crimes that are difficult to measure and poorly reported to
> police. Special studies should be conducted periodically in the
> context of the NCVS program to provide more accurate mea-
> surement of such events.**

– 4 –

Matching Design Features to Desired Goals

F ROM OUR REVIEW OF THE GOALS of the National Crime Victimization Survey (NCVS; Chapter 2) and the challenges it faces (Chapter 3), we find seven fundamental goals to have particular salience and use them as the basis for evaluating various NCVS design options. As we elaborate below, we suggest this as a set of desirable goals; they are certainly not the only possible goals, and others may place different weights on particular goal statements. Four of these goals are historical in nature, in that they reiterate or reflect the various formal task statements recounted in Chapter 2. They are:

- Production of a national measure of crime independent of official reports to the police;
- Provision of information on the context, consequences, and etiology of victimization;
- Ability to measure aspects of crime beyond the production of basic, overall rates; and
- Utility for producing information on hard-to-measure crimes that are difficult or impossible to detect in police reports.

The remaining three goals for the NCVS that we consider to be particularly relevant come from current data uses and needs:

- Capability to readily provide information on emerging crime problems;

- Capability to provide small-domain and subnational data of direct interest to states or localities; and

- Timeliness of resultant data (by which we mean that policy-relevant data can be collected and tabulated sufficiently quickly as to assess emerging trends and inform policy responses).

This chapter summarizes the relationships between the goals of the NCVS, the design of the survey, and the implications of various designs in terms of cost, error, and utility. We first consider alterations to the current NCVS design that could be put in place quickly for the purpose of saving money, while more substantial changes in design are assessed and implemented (Section 4–A). We then consider long-term changes in the form of the NCVS (4–B), outlining a set of survey design packages, some representing relatively minor changes to the current design and others overhauling the basic approach to measuring victimization. Section 4–C describes the trade-offs in cost, error, and utility associated with various design features and presents our general assessments.

4–A SHORT-TERM FIXES: COST REDUCTION STRATEGIES AND THEIR IMPLICATIONS

Any thoroughgoing redesign of the NCVS will require research and development work. At the same time, the survey is in dire straits with respect to funding. The Bureau of Justice Statistics (BJS) has managed this problem by making short-term changes in the survey that will allow it to continue while the panel does its work and while some research and development work is done. These changes are designed to reduce cost with the most minimal disruption to the survey. However, most have been implemented with no empirical tests of their likely impact, a very risky survey management strategy. In this section, we review a number of cost reduction strategies and assess their implications for the cost, error, and utility of the survey. Some of these strategies have been introduced into the survey and others have not.

Table 4-1 describes some short-term changes that might be carried out in order to reduce NCVS costs (some of which are already planned for implementation in 2007 as part of the most recent cost-saving efforts by BJS). One of these changes—reduction in sample size—has been the alternative of most frequent resort over the history of the NCVS. The table does not include a change in the reference period of the NCVS (i.e., from 6 to 12 months) even though—as we discuss in Section 4–C.1—we favor it as an alternative to continuing to reduce the NCVS sample size.

Two small-group expert meetings convened by Lauritsen (2006b:9) suggested reductions in the length of the NCVS questionnaire as a potential source of cost savings. Several of the alternatives we discuss in the next section include attempts to streamline the content of the survey; however, we do not include reductions in questionnaire length in the table of short-term alterations. As the summary of those meetings indicates, cost-saving implications from questionnaire length are not immediately obvious:

> Efforts to shorten interviewer *screening* time are likely to produce very small savings because the field costs of conducting the screening interview are likely to be the same. Eliminating or vastly reducing *incident* report details for less serious crimes may result in greater savings, especially if such experiences account for some of the field costs associated with future efforts to improve retention and participation.

In general, condensing the content of the incident report portion of the questionnaire was seen as a better alternative because "it would reduce respondent burden and perhaps minimize errors such as survey fatigue and future participation." However, we again note that large portions of the NCVS costs arise not in interviewing but in contacting and gaining the cooperation of sample households.

We think that it is critical to emphasize that even small changes to the design of a survey can have significant impacts on resulting estimates and the errors associated with them. Design changes made (or forced) in the name of fiscal expediency, without grounding in testing and evaluation, are highly inadvisable.

Recommendation 4.1: BJS should carefully study changes in the NCVS survey design before implementing them.

We use "study" in wording this recommendation because the appropriate measures may vary based on the specific methodological changes being considered. The comprehensive redesign of the NCVS included a sufficiently large bundle of changes that implementation was phased in over the course of several years; during that time, data were available using both the old and new survey instruments, for comparison and evaluation purposes. Not all possible changes would require such a lengthy and costly phase-in process, but some would; these include changes in the stratification of the sample of addresses or a shift in reference period from 6 to 12 months. A study of proposed changes may include smaller scale testing or the reanalysis and recalculation using existing data; changes in this class might include decisions on how to count series victimizations in estimates (see Section 3–B.2).

As noted in Table 4-1, one adjustment that BJS decided to implement in 2007 (for the production of 2006 NCVS estimates) was to include the first

Table 4-1 Short-Term Changes to the NCVS to Achieve Cost Reductions

Changes to Current Design	Cost	Error	Utility
Using the bounding interview in NCVS estimates*	This would reduce costs if the effective 1/6 increase in sample was offset by a sample cut. The methodological work on the adjustment would involve costs.	Some increase in standard errors due to adjustment.	Availability of unbounded interviews advantageous for methodological research.
Keeping units in sample for an eighth interview	Effectively increases sample size by 1/6, which would reduce costs if it was offset by a cut in sample.	Some increase of time in sample bias. No adjustments currently made for known time in sample bias.	None.
Including series incidents in the estimates	Reductions in standard errors due to higher victimization rates would permit sample size reductions or other cost savings. Some methodological work necessary to select maximum number of incidents to allow in a series.	Some reduction in standard errors due to the increase in the rate of victimization, especially violent victimization and the inclusion of some suspect events that cannot be accurately allocated to reporting years or crime category.	High-end users have always had access to series incidents. However, including them in the general estimates could benefit other NCVS consumers, as the products would more accurately estimate the magnitude of such incidents as domestic assaults and assaults at school.
Reduce or eliminate allocation to centralized CATI facility*	Unclear; use of centralized CATI was intended as major cost saver, but BJS and Census Bureau experience suggests no gain over current mix of in-person and informal CATI interviews.	Pilot work on CATI showed large effect on reporting, but early implementation suggested that centralized CATI use yielded higher reported incidence rates (Hubble, 1999). This seems likely to be due to the automation of the instrument, generally; estimates may not vary much between interviewer-initiated calls using a CAPI instrument and calls from a centralized CATI center.	The mode of interview would be somewhat more uniform across the sample. There might be some reduction in interview refusals due to caller ID screening.

Sample reductions	Cost savings in proportion to the size of the reduction.	Increases in standard errors of estimates commensurate to size of the reduction.	Decreasing precision of the estimates for violent crime; further reductions would seriously limit usefulness of estimates.
Biannual estimates	Current BJS reports already combine two years of data to assess short-term change; this would formalize this approach by switching to a 2-year release cycle. This would reduce costs if the increases in precision were offset with sample cuts or other cost reductions.	Aggregating two years of data would increase the precision of the estimates somewhat.	Threatens the relevance of survey as a source of annual information on crime.
Subsampling prevalent incidents for receiving incident form	This could reduce time in household but since much of the cost is in contacting households this will not save much.		
Use police statistics to improve stratification in the sample	To the extent that stratification reduced standard errors, commensurate sample cuts could reduce costs.	If stratification was used without sample cuts, then there would be gains in precision with changes in sample stratification.	This revision stops short of a wide-scale revision of survey content but is not really a short-term effort. In addition to planning, it would take several years to phase in and implement.

* Planned for implementation in 2007.

NOTE: CATI, computer-assisted telephone interviewing. CAPI, computer-assisted personal interviewing.

interview at a sample address in the estimates. The NCVS has always included some mix of bounded and unbounded interviews in its estimates due
to movers—households that depart an address during that address's 3.5-year
stay in the survey sample and are replaced with new households; the first interview with the "new" household at an "old" address cannot be bounded
by the previous interview. Catalano (2007:131–132) observes that the level
of unbounded interviews included in the NCVS estimates "has fluctuated
between 10.7 and 14.7 percentage points during the course of the survey";
Lauritsen (2005:Note 3) comments that the fraction "is roughly 6% and
does not appear to have changed much over the past decade." Bounding
had been the focus of some studies prior to the 1992 NCVS redesign (see,
e.g., Biderman and Cantor, 1984; Murphy, 1984; Woltman et al., 1984), but
there was a paucity of research on combining bounded and unbounded data
in postredesign NCVS estimates. Addington (2005) was able to assess the
relative effects of bounding and residential mobility on reported victimizations in what may be the only published, peer-reviewed article in the postredesign era. However, due to its reliance on public-use files, that analysis was
limited to data from an administration of the School Crime Supplement for
which incoming respondent (first interview) data were included. The decision to include unbounded, first interviews in NCVS estimates was made as
our panel was being established and assembled, and so we do not think it
proper to second-guess it; we understand the fiscal constraints under which
the decision was made. However, it serves as an example of a seemingly
short-term fix with major ramifications, and it would have benefited from
further study prior to implementation.[1]

4–B   ALTERNATIVES TO THE CURRENT DESIGN IN THE LONG
TERM

Tables 4-2 and 4-3 present a set of design packages that can be compared
with the current NCVS design. The columns of Table 4-2 describe a set
of specific design features on which the packages vary. Each design feature
poses some trade-off decision for BJS, offering benefits of some type, but
also potential costs or risks of not fulfilling the BJS mission with regard
to the measurement of crime. We discuss first the columns of Table 4-2,
the specific design features, by pointing out the benefits and costs of each

---

[1]On December 12, 2007, the first 2006 estimates from the NCVS were released. The
inclusion of the first-time-in-sample interviews, combined with the effects of a new sample
based on 2000 census information, led to "variation in the amount and rate of crime [that] was
too extreme to be attributed to actual year-to-year changes"; BJS concluded that "there was a
break in series between 2006 and previous years that prevented annual comparison of criminal
victimization at the national level" (Rand and Catalano, 2007:1).

feature; assessments of advantages and disadvantages of each design package follow in Table 4-3.

We emphasize—and elaborate in greater detail in Section 4–B.2—that the set of design packages we discuss in this chapter is not, and is not intended to be, exhaustive of all design possibilities for the NCVS. Instead, they are meant to suggest a range of choices and broad visions for the survey. It should also be noted that we do not have the capacity to provide an estimated price for all of these various options; while some can clearly be assumed to be less costly than the current NCVS design, others may not. Our focus is on the mapping of possible design choices to desired goals, and the weight one puts on the end cost of the design (relative to other choices) is important in selecting one; we think that it is our charge to suggest possibilities but not to impose any particular weighting of goals and design characteristics.

4–B.1 Characteristics of Possible Design Packages

Table 4-2 compares each of these alternative design packages across nine characteristics, five related to the general survey design and four specific to the type of instrumentation used to implement the design.

Single Survey Versus System of Surveys

The wealth of information needed to help policy analysts and researchers alike is ever changing. One survey—even one that is omnibus in nature—may not be able to handle changing needs. A "system of surveys" is a set of independent but coordinated data collections; in the context of victimization, the individual surveys could vary by the type of crime studied, the setting of the crime, or the population victimized. A system of surveys may be better suited to focus on measuring a particular type of crime or a particular social context; a system of surveys may also be amenable to a "quick survey" capability, being able to provide meaningful data in a compressed time period (a few weeks, rather than several months or years) in response to critical social events, such as a college or school shooting. A system-of-surveys approach could increase the richness of covariates and so may be more useful for the study of correlates of victimization. However, this greater flexibility must be reconciled with an increased burden of coordination and heightened dependence on the continued existence (and funding) of, and access to, multiple other surveys. By comparison, the single-survey approach allows for focusing of effort and consistency of measurement.

General Structure of Sample

By "general structure of sample," we mean the combination of two concepts. One is the unit of analysis used in the survey: choices include households, persons, addresses, or phone numbers. The second is the way the sample of those units is selected: for instance, an element survey makes draws directly from a frame or listing of eligible units, compared with clustered or multistage approaches. Choices for the sample design affect the efficiency of the data collection or the degree to which respondents can be found at units chosen for the sample. Due to extensive clustering, a multistage sample is less efficient than an element sample, as is the case in a pure random-digit dialing (RDD) survey. Oversampling (state, local area, or small population) improves the precision for those areas or populations in which sample is added but not the overall national picture.

Under this heading, we also include potential alternatives to the NCVS current design that would serve to boost basic efficiency. One such approach would be adding crime-related questions as supplements to existing surveys, such as the American Community Survey (ACS), National Health Interview Survey (NHIS), or Current Population Survey (CPS), all of which are conducted by Census Bureau.

Cross-Section or Panel

Panel surveys are generally useful to assess the change over time and permit longitudinal analysis of change in behavior. The NCVS is a 7-wave panel of addresses, not individuals; this permits analysis of longitudinal behavior of nonmoving households.[2] A rotating panel design also permits bounding interviews with nonmoving households, using information collected in a prior interview to make sure that events are not duplicated. Rotating panels are less costly than the repeated cross-sectional surveys and produce lower cost per interview. However, the advantage of rotating panels on response rates is likely to be a function of how frequently sample households are visited; households may drop off in their cooperation with a survey the longer they remain in the sample. By comparison, repeated cross-sectional surveys may have a lower response rate but can be more flexible in terms of content and design modifications.

---

[2]It is important to note that the NCVS panel of addresses is different from the panel of *persons* used in some other surveys, such as the Survey of Income and Program Participation and the National Longitudinal Survey of Youth. Although a panel-of-persons structure would be ideal for capturing within-household experiences of victimization (and reactions to them), a panel of addresses is a substantially less costly compromise approach.

Reference Period

The NCVS currently asks respondents to report victimizations occurring in the six months prior to the interview ("a 6 month reference period"). The current practice came about as part of the 1992 NCVS redesign and was intended as a simplification for respondents. Prior to the redesign, the survey questionnaire emphasized boundaries on the reference period, asking respondents "to report for the 6-month period ending at the end of the previous month and beginning 6 months prior" (e.g., January–July for an interview conducted in August). The new, and current, questionnaire implicitly asks respondents to report victimizations up to and including the day of the interview by referencing only the past 6 months (Cantor and Lynch, 2000:112). By comparison, most other ongoing victimization surveys (e.g., the British Crime Survey and the International Crime Victimization Survey; see Appendix E) use a 12-month reference period.

Between 1978 and 1980, the Census Bureau conducted a reference period research experiment using parts of the National Crime Survey sample; see Bushery (1981a,b); additional early work on reference period under the initial design is described by Singh (1982) and Woltman and Bushery (1984); see also the summary by Cantor and Lynch (2000).

Mode

Telephone interviewing reduces the proportion of time that interviewers are engaged in noninterviewing activities (i.e., travel). Thus, the mode tends to be cheaper per interview than face-to-face interviewing. However, the telephone model appears to be more sensitive to longer interviews, exhibiting higher break-off rates and partial interviews. Furthermore, the telephone prohibits use of visual aids in measurement, such as the flashcards currently used in the NCVS to clarify responses to some questions (e.g., race, Hispanic origin, employment, education) and provide respondents with information about survey privacy (U.S. Census Bureau, 2003:A7-6–A7-9). For some types of populations (e.g., those in multiunit structures and walled subdivisions), the telephone might yield higher contact rates, but for others there appears to be no mode difference. The face-to-face mode, while costly, generally is viewed as yielding the highest cooperation rates. The self-administered mode (e.g, paper questionnaires, web administration) is gaining increasing attraction. Self-administration often has very low per-unit costs. Self-administration is found to yield more honest reporting on sensitive topics (e.g., domestic violence, rape).

Screener and Incident Form

The current NCVS separates the functions of counting and listing victimization events from more detailed classification of those crime events through its separate screening questionnaire and incident report. This separation of screening from classification is an effective way of promoting recall and filtering out ineligible events; moreover, the additional information collected in the more detailed incident form permits a fuller description of crime events and their context. It is possible for surveys to take a different, more omnibus approach, integrating screening and classification in one pass. Surveys that use responses to screening questions to classify events may be less burdensome, but they may yield data of lower quality. (Further discussion of the nature of the screener portion of the questionnaire follows under "Event History.")

Core and Supplement

Some surveys have a set of questions that are consistently asked of all respondents, sometimes labeled the "core." The full survey questionnaire contains core questions and a rotating set of supplement questions. Scheduled supplements allow topical reports from the survey, enriching the breadth of reports. These supplements might change over time, to reflect the changing nature of crime.

Subsampling Frequent Events

In the current NCVS design, respondents are asked to complete an incident report for every victimization incident counted during the screening portion of the questionnaire. A possible approach to reduce overall interviewing time (and hence lower costs somewhat) is to subsample the most frequently occurring victimization types (e.g., theft, simple assault) and to collect the detailed incident report data for only some of these incidents. The subsampling rate could depend on the frequency with which a particular crime is reported as well as complex structures involving multiple crimes. For example, a matrix sampling framework could be used to subsample incident reports to preserve certain associations or joint occurrences. Probability sampling of such events permits estimation of standard errors of estimates. Subsampling could be implemented with slight modification of the current computer-assisted instrument used by the NCVS.

Event History

We include "event history" as a category as one relatively new technique in survey methodology that may be useful to consider in promoting accu-

racy in the screening portion of the NCVS interview, in addition to continued emphasis on completeness of recall. As described in Section 3–B.1, the
problem of respondent recall—that survey respondents often have difficulty
recalling the number or exact timing of events and that telescoping may
occur as a result—is a long-standing one in survey research. Accordingly,
a major objective of the 1992 NCVS redesign was to improve the screening portion in order to promote completeness of reporting. As Cantor and
Lynch (2005:296) observe, the structure and wording of the screener questions made the new NCVS the first federal statistical survey to be designed
based on psychological models of the survey response process, or what have
come to be known as cognitive aspects of survey measurement (see, e.g.,
National Research Council, 1984; Sudman et al., 1996; Sirken et al., 1999;
Tourangeau and McNeeley, 2003). Specifically, the new NCVS screener emphasized the cognitive steps of comprehension and retrieval, adding a large
number of "short cues" and structuring the interview into multiple frames
of reference (Cantor and Lynch, 2005:297–298). Martin et al. (1986) review screening procedures considered in the redesign. (In addition to the
revisions to the screening questions, the NCVS retained its high-level approach to dealing with recall problems by withholding the first interview
at an address and using it only to "bound" responses given in the next interview, until that practice was changed for the production of 2006 NCVS
estimates.)

In recent years, survey research has suggested methods for structuring and designing questionnaires that can improve the recall and dating of
events. These event history methods typically involve the use of calendar-
type structures in questionnaires, along with special cues, that permit respondents to make best use of thematic groupings and landmarks in the way
memories are coded in the human mind. See, for example, Belli (1998) for
an overview of event history methodology and Belli et al. (2001) for a direct
comparison between event history and standard "question list" methods.
The implementation of event history methods in the NCVS would require
major restructuring in the existing screener portion of the survey. However,
they could become particularly important if respondents were asked to recall
victimization incidents over a longer time frame, as would be the case with a
switch from a 6-month to a 12-month reference period. Although they tend
to improve recall of events, event history methods generally require more
interviewing time. They are also generally easier implement in face-to-face
interviews—a setting in which a flashcard or a questionnaire section structured in the form of a calendar grid—than in telephone interviewing, which
is the mode most commonly used in NCVS interviews. However, event history methods have been developed for use in telephone surveys; Belli et al.
(2007) compare the implementation of an event history approach for telephone interviews in the Panel Survey of Income Dynamics with standard

telephone interviewing methods and find benefits in improved retrospective reporting. The implementation of event history techniques in the NCVS may improve the accuracy of the data but may also increase the cost, and so would need to weighed and dealt with in the context of other modifications adopted by the NCVS.

4–B.2   Survey Design Packages for Comparison

There are many possible combinations of specific survey design features that satisfy the different characteristics that form the columns of Table 4-2. We focus on 10 survey design bundles that, in the panel's view, contain design features that are compatible and form a cohesive design option. Several of them are modeled after data collection efforts already in place, whether in other countries' victimization surveys or in other social surveys of the American public.

The rows of Table 4-2 describe a set of alternative structures for the NCVS, beginning (with item (0)) with the current design.

Core-Supplement Models (1)–(2)

Core-supplement models would retain the rotating panel design of the current NCVS (contacting the same households for several interviews) but differ from the current NCVS in the emphasis placed on topical supplements. They are a partial implementation of the British Crime Survey (BCS) model in that they would make topical supplements a regular, built-in part of the survey structure (paired with a streamlined core NCVS questionnaire) instead of being conducted on an irregular and as-available basis. We differentiate between (1) a core-supplement model that would maintain the 6-month reference period of the existing NCVS for maximum continuity of estimates and (2) a model that would shift to a 1-year reference period (and fewer interviews with the same household).

British Crime Survey Type (3)

The British Crime Survey type model (3) is built around a core-supplement design but would represent a major shift from the current NCVS as it would abandon the rotating panel design for an annual cross-section sample. It would also potentially include the addition of event history methods to improve recall of victimization incidents. This option is described as "BCS Type" rather than a pure replication of the BCS design, because some key parts of the design are left unspecified (and subject to change). Specifically, decisions under this model type would have to be made on whether the survey is administered by a government agency or by a private firm (the BCS is contracted out) and whether interviews are collected continuously throughout the year or targeted at one specific time interval (the BCS, like the current NCVS, uses continuous interviewing).

Local-Area Boost Models (4)–(6)

Local-area boost models would maintain the existing NCVS structure but would oversample some areas or populations on a rolling basis so that subnational estimates could be produced. Given the sample sizes involved, these subnational estimates would most likely be multiyear averages of the sort produced by the Census Bureau's American Community Survey; the ACS combines data for 36 and 60 months of data collection in order to produce estimates for geographic areas as small as census tracts. In the table, we differentiate between (5) a local-area boost that would focus on states as the natural unit of interest and (6) one that would permit more flexible allocation of rolling sample, perhaps to generate estimates for large metropolitan statistical areas.

International Crime Victimization Survey Type (7)

While packages (0)–(6) generally maintain the current structure of the NCVS, packages (7)–(10) are substantially more aggressive redesigns that would effectively end the NCVS as it is currently known.

The International Crime Victimization Survey (ICVS) is an ongoing semi-annual survey administered by the United Nations Interregional Crime and Justice Research Institute. Its methodology is described in more detail in Section E–2 in Appendix E. Relative to the current NCVS design, an ICVS-type model (4) involves at least two major compromises: it would abandon both the rotating panel design and the potential face-to-face interviewing capability of the current NCVS, focusing instead on a cross-sectional telephone-only survey.[3] Replication of the ICVS one-stage screening process would also probably substantially reduce the quality of reporting in the survey. However, the instrument for such a design could fairly readily accommodate topical supplements. The ICVS-type model we describe here is not a pure replication of the ICVS, as that would abandon annual victimization measures (the ICVS is conducted on a periodic basis only, about every 4–5 years).

Partnership Model (8)

The option that we call the "partnership model" would diffuse various topics into the current NCVS into different data collection vehicles.

---

[3]As described in Section E–2, personal interviewing is carried out on a small scale in a few ICVS participant countries, and then only in the country's capital city.

Table 4-2 Current and Possible Alternative Designs for the NCVS

Description	Survey Design					Instrumentation			
	Single Survey vs System	Sample	Cross-Section or Panel	Reference Period	Mode	Screener Incident Form	Core and Supplement	Sub-Sampling Frequent Events	Event History
(0) *Current NCVS Design*	Single	Household multistage cluster	Rotating panel	6 months	CATI/CAPI	Yes	No	No	No
Core-Supplement Models—streamlined core questionnaire with structured topical supplements									
(1) *Simple*	Single	Household multistage cluster	Rotating panel	6 months	CATI/CAPI	Yes	Yes	Yes	No
(2) *Longer Reference*	Single	Household multistage cluster	Rotating panel	12 months	CATI/CAPI	Yes	Yes	Yes	No
(3) *BCS Type*—Cross-sectional design with a 12-month reference period, event history methods, routine supplements	Single	Household multistage cluster	Repeated cross-section	12 months	CAPI	Yes	Yes	Yes	Yes
Local-Area Boost Models—Cross-sectional design with a 12-month reference period, event history methods, and routine supplements, plus rolling sample to support subnational estimates									
(4) *State Boost*—Periodic increase of sample to tiers of states	Single	Household multistage cluster sample clustered by states with states rotating in and out	Cross-section	12 months	CAPI	Yes	No	Yes	Yes
(5) *Other-Area Boost*—Periodic sample increases in some other subnational groups (e.g., SMSAs or agglomerations like "rural states")	Single	Household multistage cluster sample; additional sample allocated to subnational areas rotating in and out	Cross-section	12 months	CAPI	Yes	No	Yes	Yes
(6) *Boost with Supplements*—Cross-sectional design with a 12-month reference period, event history methods, core and supplement format, a quick survey capability and a sample clustered to provide state-level estimates for rotating group of states	Single	Household multistage cluster sample clustered by states with states rotating in and out	Cross-section	12 months	CAPI	Yes	Yes	Yes	Yes

Description									
(7) *ICVS Type*—Pure ICVS replication would be conducted every 4–5 years and use both 5-year and 1-year reference periods	Single	Households; RDD	Repeated cross-section	12 months/ 5 years	CATI	No	Yes	No	No
(8) *Partnership Model*—System of victimization surveys with a supplement to an ongoing household survey and a series of interrelated surveys focused on particular crimes or populations	System	1) add current screener-type questions for crimes well-reported to police (e.g., robbery, burglary, motor vehicle theft) to other national survey vehicle, like ACS or CPS; 2) routine surveys of domestic violence, workplace, and schools, possibly including general "harm" survey to include assault; 3) episodic special topic surveys on looming issues	Either	Varied	Mixed	Yes	No	NA	NA
(9) *Surveillance Model*—Distributed system of surveys, administered by central core but conducted through BJA-type grants to state/university consortia	System	1) Individual state surveys would follow core content, with annual appropriations 2) NCVS program would administer content. Ideally, there would be a continuing, small-level national sample for calibration; a pure version of this model would make the "NCVS" strictly the compilation of the state-level surveys	Either	—	Mixed	Yes	—	No	—
(10) *Periodic Survey* without incident reports: Cross-sectional design with changing annual content based on policy interests; not expressly limited to victimization but could include questions on perception of risk	Single	Phone (likely RDD)	Cross-section	—	Mixed	No	No	No	No

NOTES: ACS, American Community Survey. BCS, British Crime Survey. BJA, Bureau of Justice Assistance. CAPI, computer-assisted personal interviewing. CATI, computer-assisted telephone interviewing. CPS, Current Population Survey. ICVS, International Crime Victimization Survey. RDD, random-digit dialing. SMSA, standard metropolitan statistical area.

A basic premise of this model is that it would still be useful to have some measure—independent of the police—of certain crime types that are generally well reported to police, such as robbery, burglary, and motor vehicle theft, but that it might not be necessary to gather specific incident-level data on these crimes. Hence, the partnership model would shift the basic screener questions for these crime types to some other national survey—possibly the American Community Survey (ACS) or the Current Population Survey (CPS), both conducted by the Census Bureau—and waive the collection of detailed incident data for them.

Having shifted measures of some crimes to other surveys, a series of smaller scale, interrelated surveys could be targeted to specific crimes or populations. This would have the benefit of training resources on crime types for which a sample survey is best suited (relative to reliance on administrative or facility records). For example, some continuity in the measure of assault could be preserved though a smaller scale harm survey, perhaps combining information from a smaller scale survey with emergency room data. Although the choice to focus efforts on individual surveys for specific crime types would ideally be driven by the desire to increase the quality of information, it could also come about as the result of cost cutting. Hence, the number and frequency of these smaller surveys—and the fiscal resources dedicated to them—would be a major consideration.

Surveillance Model (9)

While the partnership model differs by spreading the NCVS topic area content over a number of alternative sources, the surveillance model (9) maintains fairly tight control over topic area content but disperses the data collection task. This model is based mainly on the Centers for Disease Control and Prevention's (CDC) Behavioral Risk Factor Surveillance System (BRFSS) (described more fully in Box 4-1), in which CDC provides funding to state health departments to collect data. It is also based in part on the Federal-State Cooperative Program for Population Estimates (FSCPE) between the Census Bureau and state demographic units, which has become an important partnership agreement for improvements in various population estimate and general census processes.

Most significantly, though, we consider this model because of BJS's unique placement relative to other federal statistical agencies: inside the Office of Justice Programs, whose core mission is providing assistance to state and local law enforcement agencies. Under the BRFSS, CDC provides core funding and central coordination to state health departments to administer a telephone survey. Likewise, under this model, BJS and the Bureau of Justice Assistance would provide funding and central support to states (or consortia of states) to collect victimization data, either in-house or through subcontract with private or university survey research organizations. In return for administering a core set of questions and, optionally, topic supplements, states could add their own questions to the survey. As a state-based system, the states (or groups of states) would also have direct estimates at their level of geography. If this network of affiliates were to provide full coverage of the nation, the state-level data could be pooled to provide a national-level estimate.

Crime Poll (10)

A final option follows the example of the Texas Crime Poll, a semiannual data collection that is directly intended to inform specific legislative issues at the state level; the poll is described in more detail in Box 4-2. Alternately, it can be viewed as promoting a set of questions that have historically been handled through supplements to the NCVS—questions on popular attitudes on the extent of crime and perceptions of public safety—to be the exclusive contact of the victimization survey.

This option is the starkest contrast with the current NCVS design, as it would involve giving up the goal of producing annual rates of criminal victimization and the collection of detailed incident information. However, it is also—almost certainly—the lowest cost alternative considered in this table. Depending on its implementation and how close it follows the Texas example, it could also arguably be the most responsive to specific legislative directives, the least burdensome on individual respondents, and would be readily amenable to questions on attitudes on crime and justice issues other than victimization. The specific implementation envisioned by the model would be a representative cross-sectional survey, conducted by telephone.

Other Design Possibilities

We do not intend these 10 design packages to be exhaustive of all alternative design possibilities; rather, these 10 are chosen to illustrate a range of design choices. In our judgment, they constitute packages that merit first review, either because they force attention to their impact on alternative goals of the NCVS (and thus sharpen thinking on goals) or focus attention on innovations that could reduce the costs or improve the measurement efficiency of the NCVS (and thus focus attention on process). We do not wish to exclude other combinations of design features; indeed, detailed consideration of the 10 options may identify a combination of design features that is preferable to any of the 10 packages identified here. Thus, the 10 options are a first step in a decision process to choose which of the goals of the NCVS should be emphasized and which should be deemphasized in the future.

Box 4-1   Behavioral Risk Factor Surveillance System

The Behavioral Risk Factor Surveillance System (BRFSS) is a collaborative data collection project in which data are collected by state health departments through monthly telephone interviews, with funding and technical and methodological coordination from the Behavioral Surveillance Branch of the Centers for Disease Control and Prevention (CDC). Initiated in 1984 with 15 participating states, the BRFSS had participation by all 50 states, the District of Columbia, Puerto Rico, Guam, and the U.S. Virgin Islands as of 2001. The system is intended to measure risk behaviors and preventive practices in the adult population (18 years of age or older).

The BRFSS has a set of core questions that are agreed to by the states and CDC and that are administered in all states without modification. Since a 1993 redesign, some of these core questions are fixed and asked every year while a "rotating core" set of questions is asked every other year. In addition, some slots on the core questionnaire are reserved for "emerging issues questions." The individual states may also elect to add optional topical modules that are developed by CDC; optional modules were first fielded in 1988. Other branches within CDC sponsor questions in the BRFSS, as do external agencies such as the National Institutes of Health and the Department of Veterans Affairs. States are also permitted (and encouraged) to add their own questions to the instrument without direct CDC editing; the state health departments may obtain funding from other state agencies to place questions on the BRFSS questionnaire.

BRFSS samples are drawn by CDC from a commercial telephone number database, or they may be constructed by the states on their own as long as they comply with general standards. CDC's objective is to support at least 4,000 interviews per state, although effective sample sizes vary; in 1995, Iachan et al. (2001:Table 1) report that state sample sizes ranged from 1,193 (Montana) to 5,107 (Maryland). Most state health departments let contracts to commercial or university survey research units to conduct BRFSS interviews; in 2006, only 14 state health departments conducted the interviews in-house.

The BRFSS is intended to generate state-level estimates, and over time the BRFSS has sought to provide both higher and lower level estimates. As an amalgam of state samples, the BRFSS is not intended to provide direct national estimates; "nevertheless," Iachan et al. (2001:221) observe, "there is much interest among the research community in using the BRFSS for such estimates because all states use the same core instrument, sample size is relatively large, and the annual data are available within 6 months after collection." Iachan et al. (2001) discuss different methods for pooling the state samples and find consistent results between their national BRFSS estimates for some items with corresponding items from the National Health Interview Survey. In 1997, pressures for estimates below the state level resulted in the Selected Cities Project and, since 2002, the Seletcted Metropolitan/Micropolitan Area Risk Trends (SMART) program. Through the project, estimates have been generated for counties and metropolitan/micropolitan statistical areas in which at least 500 BRFSS interviews are completed in a year; estimates were available for 99 cities in 2000, and SMART estimates were derived for 145 statistical areas and 245 counties in 2006.

Box 4-1   (continued)

The BRFSS has been used to collect data on some incidents of violence. A module of questions on intimate partner violence and injuries was implemented in the state of Washington in 1998 (Bensley et al., 2000). In 2005 and 2006, CDC offered both an 8-question sexual violence module and a 7-question intimate partner violence module; 29 states or territories opted to use one or both of these modules during 2005–2006. Individual states have periodically added their own questions on violence-related matters, including generic questions on injury and on access to firearms.

SOURCES: Centers for Disease Control and Prevention (2003, 2005, 2006a,b, 2007).

Other design features could be explored or adjusted to forge additional design options:

- Variations in the number of repeated interviews conducted at a sample address (and, correspondingly, the length of time an address remains in sample) can be made within the framework of a rotating panel design as is currently used in the NCVS. Although the nature of the current NCVS as a longitudinal sample of addresses is one of its key attributes and has provided the basis for bounding interviews, that longitudinal structure is an underutilized feature. NCVS data files have not commonly been constructed in linked, multiyear longitudinal segments; most recently, this was done for 1995–1999 data (Bureau of Justice Statistics, 2006b). Some of the designs we consider in Table 4-2 would switch from a rotating panel to a cross-sectional design, but it is also possible to conceive designs that would emphasize longitudinal structure (e.g., conducting—in part or in whole—a longitudinal sample of people or households, making efforts to contact the same respondents regardless of whether they move.)

- Implicit in what we call the partnership model (8) is the notion of shifting NCVS screener-type questions to another survey, like the American Community Survey (ACS) or Current Population Survey (CPS). This is one way in which an alternative mode of survey collection— via the mail—could be used in victimization framework, but not the only one. Particularly if the NCVS continues to be collected by the Census Bureau, one approach could be to use the ACS or the CPS as a prescreener: construction of the NCVS sample could be targeted based on "yes" or "no" responses to victimization questions on the other survey. (Obviously, the total burden on respondents who would then be included in multiple federal surveys would have to be taken into account.) More generally, it is possible to consider detaching the NCVS screening questionnaire by mail and then performing follow-up by phone or personal interview (although this would hurt the cuing

Box 4-2   Texas Crime Poll

First conducted in 1977, the Texas Crime Poll is a semiannual public opinion survey administered by the Survey Research Center, Criminal Justice Center, Sam Houston State University. The poll is authorized by the 1965 legislation that created the Criminal Justice Center, which required the center to conduct "surveys of pertinent problems in the field of crime, delinquency and corrections" (Teske and Lowell, 1977:1).

By design, "the format [of the poll] remains the same each time and many of the questions are replicated on a regular basis in order to allow for measurement of changes in public opinion. Other items are topical and are included only once" (Teske and Lowell, 1977:1). Basic attitudinal questions (e.g., 1977's "Over the past three years, do you feel the crime problem in *your* community is: • Getting better • About the same • Getting worse?") are among the questions that are repeated regularly. Topics covered by the poll questions have ranged from the appropriateness of capital punishment to perceptions of obscenity to trust in the state's use of forensic science in prosecutions. Because the poll is meant to help inform state legislative issues, question text can become detailed and complex. For instance, question 1.7 on the 2007 poll read:

> Exempting criminal justice professionals from tuition and fees would cost some universities as much as $500,000 per year. If you responded "Yes" to either items 1.5 or 1.6 above, which of the following best represents your thoughts on how these losses should be addressed? (Select only ONE of the following.)
>
> - Additional legislation should be passed to compensate the universities for these losses.
> - Tuition and fees for other students should be increased to compensate the universities for these losses.
> - The universities should consider these "losses" as their contributions to the development of criminal justice professionals and adjust their budgets to accommodate them.
> - Other options: Please specify: *Area for free response*

The Texas Crime Poll was originally conducted as a systematic random sample from the frame of persons with valid Texas driver's licenses, collected by mail; in 1977, the poll included 642 respondents (a 67 percent response rate). As of 2007, the poll remains a mail survey but its frame is based on "white pages" telephone listings. From these listings, the sample is drawn systematically, stratified by county. In its two most recent iterations (2004 and 2007), the sample design was altered in an attempt to increase the representation of minorities in the sample. In 2007, the poll included two sample groups of 1,500 people: one intended to be generally representative of the whole population and the other drawn specifically from those area code listings corresponding to areas with high-density Hispanic concentrations in the 2000 census. Respondents also were encouraged—in a follow-up postcard—to complete an Internet version of the questionnaire. However, the 2007 poll received only 332 total responses out of 2,874 valid questionnaires (11.6 percent), a decline in response rate from 2004's 22.8 percent.

Documentation and data sets for the Texas Crime Poll are archived at http://www.cjcenter.org/cjcenter/research/srp/txpi.html.

and recall structures built into the current structure, doing both the screener and incident form in the same administration).

The panel thinks that deliberate and rigorous evaluation of the design packages listed above will generate other ideas for packages that balance costs of the NCVS and the quality of key NCVS estimates. Such evaluation requires leadership within BJS and the statistical system to assemble the support for professional review of the key data series. Quick decisions in the absence of evaluation carry with them great risks of misleading the country with error-filled estimates. A careful examination is required and consistent with the principles of a federal statistical agency.

4–C ASSESSMENTS OF DESIGN FEATURES AND PACKAGES

Table 4-3 summarizes our basic assessment of how well the design packages described by Table 4-2 satisfy seven desirable goals, as articulated at the outset of this chapter. Table 4-3 also lists basic advantages and disadvantages of each design.

A basic observation from Table 4-3 is that our assessments in the table suggest an underlying difficulty concerning the goal of timeliness of the resultant data. An ideal design with respect to timeliness is one that produces estimates that can shed light on new and emerging crime trends (and types) and inform policy strategies; the design should be nimble, with the capacity to add new questions in a relatively short amount of time in order to provide "quick survey" empirical data on emerging issues. Yet we judge even the more streamlined core-supplement models as being ill-suited to this ideal timeliness goal; the only design that meets the goal adequately is the crime poll (10) design in which "quick survey" capability is the entire measurement goal.

Furthermore, the table suggests two other, related conclusions that deserve emphasis because they shape the discussion that follows. The first is that there is no especial magic to any of the profiled designs: none of the listed design packages is optimal across all of the desirable goals, and each represents certain trade-offs and compromises. As we note below, there are specific design elements that we think are worthy of consideration in the NCVS, but we do not think that any of the full-blown design packages—as a whole—is uniquely superior to the others. The superiority of packages depends on one's choices of goals and the weight that is placed on them; hence, scoring the alternative models and picking a "best" one is not as simple as counting the numbers of shaded boxes in the table. We do not presume that our set of listed goals is uniquely correct, or that other stakeholders would provide exactly the same assessments as we do, and so do not intend to try to justify a single package as better than all of its competitors.

Table 4-3 Goals of the NCVS and Alternatives to the Current Designs

● indicates that the panel believes that the goal is *well* served by the design package; ◐ that the goal is *adequately* served; ☒ that the goal is *poorly* served.

Design and Products	Nat'l Measure Indep. of Police	Vict. Context and Etiology	Emerging Crime Problems	Detail for States and Localities	Timely	Info Beyond Crime Rates	Hard-to-Measure Crimes	Pros	Cons
Current NCVS • Annual estimates of level and change • Special reports on social context of crime and on specific types of crime and subpopulations of victims	●	●	×	×	×	●	●	• High-quality national estimates of rates for "street crimes" • Long, stable time series • DoJ controls survey content • Omnibus vehicle reserves resources for victimization data • Limited coordination with other agencies or entities required	• Limited detail on crime events • Inflexible information content • Not particularly timely • Scope of basic BJS reports restricted to rate estimation, though data have been used extensively in other ways • Expensive • Little information of direct use to states and localities
Core-Supplement (simple) • Annual estimates of level and change • Regular series of topical reports based on scheduled supplements	●	●	●	×	×	●	●	• DoJ controls survey content • Omnibus vehicle reserves resources for victimization data • High-quality national estimates of rates for street crimes • Long, stable time series for core items • More information on context, etiology, and consequences • More information on emerging crime problems • More information on issues beyond crime rates • Reduced cost from trimming core	• Development and fielding of supplements will add to survey cost • Some increases in coordination with other agencies • Additional staffing for BJS • Little information of direct use to states and localities

Table 4-3 (continued)

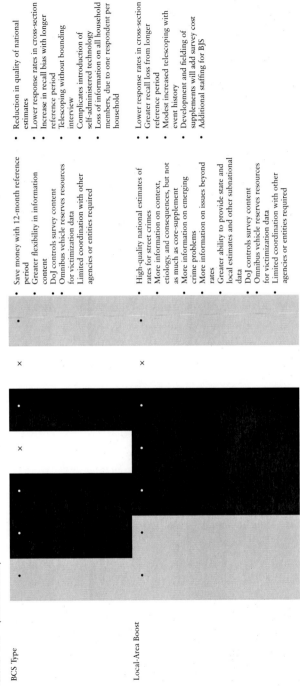

BCS Type			

BCS Type

- Save money with 12-month reference period
- Greater flexibility in information content
- DoJ controls survey content
- Omnibus vehicle reserves resources for victimization data
- Limited coordination with other agencies or entities required

- Reduction in quality of national estimates
- Lower response rates in cross-section
- Increase in recall bias with longer reference period
- Telescoping without bounding interview
- Complicates introduction of self-administered technology
- Loss of information on all household members, due to one respondent per household

Local-Area Boost

- High-quality national estimates of rates for street crimes
- More information on context, etiology, and consequences, but not as much as core-supplement
- More information on emerging crime problems
- More information on issues beyond rates
- Greater ability to provide state and local estimates and other subnational data
- DoJ controls survey content
- Omnibus vehicle reserves resources for victimization data
- Limited coordination with other agencies or entities required

- Lower response rates in cross-section
- Greater recall loss from longer reference period
- Modest increased telescoping with event history
- Development and fielding of supplements will add survey cost
- Additional staffing for BJS

Table 4-3 (continued)

Design and Products	Nat'l Measure Indep. of Police	Vict. Context and Etiology	Emerging Crime Problems	Detail for States and Localities	Info Beyond Crime Rates	Timely	Hard-to-Measure Crimes	Pros	Cons
Boost with Supplements	×	× •	•	•	•	×	•	• High-quality national estimates for rates of street crimes • More information on context, etiology, and consequences but not as much as core-supplement • More information on emerging crime problems • More information on issues beyond rates • Greater ability to provide state and local estimates and other subnational data • DoJ controls survey content • Omnibus vehicle reserves resources for victimization data • Limited coordination with other agencies or entities required	• Some reductions in quality of information • Lower response rates in cross-section • Greater recall loss from longer reference period • Modest increase in telescoping with cross-section and event history • Development and fielding of supplements will add survey cost • Additional staffing for BJS
ICVS Type	×	×	•	×	•	×	×	• Reduced costs of sampling and administration with all-telephone RDD design • Changing information content is simple • DoJ controls survey content • Limited coordination with other agencies or entities required	• Large reductions in the quality of the data • Much lower response rates • Increased recall bias due to longer reference period • Underreporting due to abandonment of screener/incident logic • Increased telescoping • Increase in out-of-scope events due to lack of incident form • Loss of all geographical information • Loss of detail incident information • Reduced resource set devoted to victim statistics

Table 4-3 (continued)

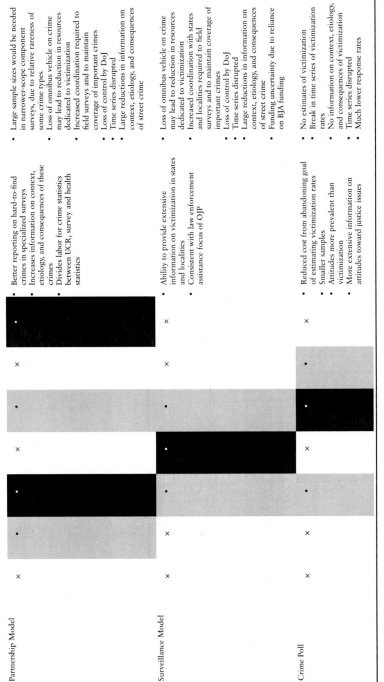

Partnership Model

- Better reporting on hard-to-find crimes in specialized surveys
- Increases information on context, etiology, and consequences of these crimes
- Divides labor for crime statistics between UCR, survey and health statistics

- Large sample sizes would be needed in narrower-scope component surveys, due to relative rareness of some crime types
- Loss of omnibus vehicle on crime may lead to reduction in resources dedicated to victimization
- Increased coordination required to field surveys and to maintain coverage of important crimes
- Loss of control by DoJ
- Time series disrupted
- Large reductions in information on context, etiology, and consequences of street crime

Surveillance Model

- Ability to provide extensive information on victimization in states and localities
- Consistent with law enforcement assistance focus of OJP

- Loss of omnibus vehicle on crime may lead to reduction in resources dedicated to victimization
- Increased coordination with states and localities required to field surveys and to maintain coverage of important crimes
- Loss of control by DoJ
- Time series disrupted
- Large reductions in information on context, etiology, and consequences of street crime
- Funding uncertainty due to reliance on BJA funding

Crime Poll

- Reduced cost from abandoning goal of estimating victimization rates
- Smaller samples
- Attitudes more prevalent than victimization
- More extensive information on attitudes toward justice issues

- No estimates of victimization
- Break in time series of victimization rates
- No information on context, etiology, and consequences of victimization
- Time series disrupted
- Much lower response rates

NOTES: ■ indicates that the panel believes that the goal is *well* served by the design package; ▪ that the goal is *adequately* served; ☒ that the goal is *poorly* served. BJS, Bureau of Justice Statistics. DoJ, U.S. Department of Justice. RDD, random-digit dialing.

The second basic conclusion is that there is nothing inherently wrong or lacking in the current NCVS design. As noted elsewhere, the NCVS has benefited from years of experience and methodological probing, particularly the intensive redesign effort that culminated in 1992. The NCVS design is a model that has been adopted by international victimization surveys as well as subnational surveys in the United States, and it is a good and useful exemplar. The principal fault of the current NCVS design is not a design flaw or methodological deficiency, or even that the design inherently costs too much to sustain, but rather—simply—that it costs more than has been tenable under current federal budgetary priorities. We argue in Chapter 3 that collection of victimization data is substantially undervalued in the United States.

There are design packages suggested in our tables that would ultimately be cheaper than the current NCVS, but they can involve considerable compromises in the quality and detail of knowledge that the present NCVS has been able to provide about crime. For instance, the partnership model (7) that would dissolve the current NCVS and allocate topics to other survey vehicles would be likely to be less expensive in the long run than the current NCVS (depending on the choice of topics for smaller, focused surveys). However, this model would involve "piggybacking" some questions (for generation of basic victimization rates) onto some omnibus survey vehicle like the American Community Survey or Current Population Survey. This piggybacking raises a number of challenges:

- By design, the NCVS screening interview is long, in order to more completely elicit victimization incidents from respondents. For consistency with other ACS/CPS questions—and to keep respondent burden under control—any victimization questions added to another survey would almost certainly have to be radically simplified, with a corresponding lack of accuracy.

- NCVS screener questions are asked of all persons in the household over age 12, in turn, whereas other surveys like CPS have only one respondent per household. That single respondent may then provide information about other household members or about characteristics of the household as a whole. Separate interviews with multiple household members would be a major departure from usual procedure for the ACS and the CPS, while reliance on proxy information from a single household respondent will lead to underreporting of victimization incidents.

Likewise, the crime poll (10) option is arguably the least expensive of the options in the table, but the sacrifices in information would be profound. Although important information about attitudes toward crime and public safety would flow readily from this type of model, there would be no capacity to obtain details about the characteristics of specific incidents.

For continuity of measurement, the best-case situation would be the provision of budgetary resources sufficient to keep the current NCVS in operation at a stable size. For long-term viability of the NCVS, the ideal would be an increase in budgetary resources—and sample size—so that at least some subnational (e.g., large metropolitan statistical areas) can be made available on some regular basis. Barring those very optimistic outcomes, we also recognize that—in terms of its sample size—the current NCVS is at a critical point: its sample size has slipped sufficiently that terminating the survey would effectively be preferable to sustaining additional across-the-board cuts in sample size.

4–C.1 Length of Reference Period

In light of these arguments, we suggest switching to a 12-month reference period (thus achieving savings by reducing the number of contacts with sample households) in the event that resources to continue the NCVS using the current 6-month reference period are not available.

> *Recommendation 4.2:* **Changing from a 6-month reference period to a 12-month reference period has the potential for improving the precision per-unit cost in the NCVS framework, but the extent of loss of measurement quality is not clear from existing research based on the post-1992-redesign NCVS instrument. BJS should sponsor additional research—involving both experimentation as well as analysis of the timing of events in extant data—to inform this trade-off.**

If this recommendation is followed, a decision will have to be made about what number of waves (interviews conducted at a particular address) would best balance cost, measurement error, and nonresponse error.

Relative to other countries' victimization surveys, the NCVS 6-month reference period is something of a luxury, with a 12-month reference period being the norm. It bears repeating, though, that the switch would decrease the cost of the survey but would not be without consequences.

The choice of a best-length reference period for the NCVS (e.g., 3, 6, or 12 months) has been a source of discussion and research since the survey's inception; tests varying reference periods were recommended by National Research Council (1976b:68), and Cantor and Lynch (2000:110–112) summarize reference period experiments conducted following that recommendation. The effects of length of reference period have also been treated in general survey methodological research. The results of these studies, NCVS-specific and otherwise, are generally quite consistent. Cognitively, an operative phenomenon is telescoping: distant events tend to be moved forward in time relative to when they occurred, while recent events tend to be shifted to an earlier date. Traumatic distant events (like victimizations) may be more easily recalled and subject to more severe forward telescoping (Bradburn et al., 1994). Accordingly—until adjustments in 2007 led to first-time-in-sample interviews being included in estimation—the first interview with an NCVS household has historically been withheld and used as a bound—a check that events are not projected forward into the 6-month reporting time window. A switch to a 12-month window means that forward telescoping might be less problematic but that complementary problems would become prominent: backward telescoping victimization incidents so that they are outside of the 12-month recall period or, for some less traumatic incident types, simply forgetting that they occurred within the past year. The extent of underreporting may differ by crime; studies like those summarized by Cantor and Lynch (2000) suggest about a 30 percent general reduction in reporting by doubling the length of the reference period to 12 months.

Consequently, it is important that a move toward a 12-month reference period be paired with research on developing event history techniques and other methods for improving accurate recall over a lengthy time window. This could involve major redesign so that events are placed on calendars or that the list of recalled victimization incidents is anchored to important dates and events (such as birthdays or anniversaries).

At the same time that the screener-type questions are revised to promote more accurate recall, attention would also have to be paid to protocols for the detailed incident form. Even allowing for the possibility that some less recent (and less traumatic) incidents may be forgotten, the net number of incidents experienced by a sample interviewee can be expected to be larger for a 12-month window than a 6-month window. Hence, respondent burden would correspondingly increase if each and every incident was subject to the detailed incident form questioning; break-offs in the interview and general nonresponse could be expected to increase as a result. Accordingly, protocols for sampling which events elicited in an interview are subject to the full incident form questionnaire would have to be tested and evaluated.

The change in reference period will also have an impact on the variances of some statistics that involve multiple incidences because of increased intraperson correlation.

Due to these logistical and technical complications, our recommendation concerning a change in reference period is deliberately nuanced. It is not a change that should be rushed into, in the name of fiscal savings, but rather one that needs grounding in pilot work and testing.

Changing to a 12-month reference period while maintaining continuous interviewing in the field (as the Census Bureau currently does) throughout the year would also warrant reconsideration of the production cycle for annual NCVS estimates. Table C-2 in Appendix C illustrates the current time coverage of events using the NCVS 6-month reference period, illustrating the distinction between collection year and data year estimates. Seeking "to publish more timely estimates from the survey," BJS switched to collection-year estimates in 1996 (the actual data release cycles have since been synchronized so that both NCVS and Uniform Crime Reports estimates of crime are made public each September). The change to collection year estimates rather than data year estimates was discussed in Bureau of Justice Statistics (2000:163), acknowledging that these "estimates for any given year will include some incidents that actually took place in the previous calendar year and will exclude some incidents that would have been reported in interviews conducted in the following calendar year." In support of that decision, collection year and data year estimates for 1995 were compared, disaggregating by type of crime, and no statistically significant differences were detected (Bureau of Justice Statistics, 2000:Appendix Table 1). Extending Table C-2 to reflect a 12-month reference period, a straight collection year estimate for year t would more thoroughly overlap events that actually occurred in calendar $t - 1$. Labeling these as "year t estimates" would thus be inherently misleading, conceptually; presenting them as estimates "for the past 12 months of collection" (and avoiding explicit mention) could be confusing to the media, the public, and other users of the data. However, reverting to a data year estimate for calendar year t would require waiting until interviews were completed in December $t + 1$ (and then allowing time for processing), which would have serious implications for the timeliness of the data.

The BCS-type model (3) described in our tables would combine a switch to a 12-month reference period with another major change: converting from a rotating panel, multiple-interview-per-household design to a cross-sectional single-interview-per-household structure. (The model also assumes that some of the cost savings would be redirected into the fielding of regular, systematic supplements, providing additional flexibility in measurement.) To be clear, we do not recommend a change to a cross-sectional design. However, it is worth noting that a 1-year reference period and cross-sectional design *could* be combined with a third major change to interesting effect. Specifically, continuous interviewing throughout the year could be replaced with focused, intensive data collection in the first few months of year $t + 1$. Doing so, respondents would have a very natural common sense reference period to work with (calendar year t), and the resulting estimates would likewise be easy to interpret as "year t" estimates. Logistically, the drawback to this scenario is the strain on field operations that would result. Prior to its switch to annual operations, the BCS was collected on this basis and the workload required the use of multiple private survey groups to carry out the interviewing workload. Such a large once-a-year effort would be a major change from current NCVS operations by the Census Bureau and would be unique among the Census Bureau's demographic survey programs. As we discuss further in the next chapter, exactly how other survey research organizations might be able to carry out the work as the NCVS data collection agent is an open question. It is also likely that the cost of moving away from continuous data collection would be prohibitive since the advantage of having a survey infrastructure in place would be lost (and such infrastructure would have to be stopped and rebuilt annually). In addition, this change in cycle would raise the problem of seasonality: the volume and type of crime varies by the month of the year. Hence, seasonality and respondent bias toward reporting more recent events could combine to distort the overall picture of crime.

4–C.2 Role of Supplements

Table 4-3 gives generally high marks to models that emphasize the role of topic supplements. Core-supplement models would streamline the body of the NCVS questionnaire to a minimal "core" and structure the remainder of the survey around regular topic supplements. Likewise, although we argue against the BCS-type model in the previous section due to a reluctance to switch away from a rotating panel design, we do find much to admire in the BCS concentration on supplements as a regular and structural part of the broader survey.

> *Recommendation 4.3:* **BJS should make supplements a regular feature of the NCVS. Procedures should be developed for soliciting ideas for supplements from outside BJS and for evaluating these supplements for inclusion in the survey.**

The role of supplements in the NCVS—as is part of the current design of the British Crime Survey (see Appendix E)—is to enhance the flexibility of the survey content. It is meant to prevent the overall content of the survey from becoming stale; in some cases, a topical supplement could serve as a "methods panel" for testing new questions that might, in time, be added to the core NCVS content. We think that the implementation of such NCVS topic supplements as the Police-Public Contact Survey and the School Crime Survey have broadened the scope of the victimization survey and that further branching out into topic supplements will serve to firm up the constituencies for the NCVS and other BJS products. Although a focus on a systematic set of topic supplements would, on its own, fall short of correcting a long-standing shortfall of the NCVS—the lack of geographic detail in estimates—it could be the vehicle for much richer substantive detail.

Recommendation 4.4: BJS should maintain the core set of screening questions in the NCVS but should consider streamlining the incident form (either by eliminating items or by changing their periodicity).

To be clear, we do not suggest trimming the screening questions on the NCVS questionnaire. As noted in Section 4–A, reducing the screener portion of the interview could slightly decrease interview length and yield very small cost savings, but the quality of resulting data would suffer. That said, we have not comprehensively reviewed the incident form to suggest items to cut, either. The point of this recommendation is that if reductions in interview length are deemed necessary (particularly in order to facilitate a more regular set of topical supplements), then some means of condensing incident form content should be considered.

It follows from this recommendation that we think that BJS and the Department of Justice should dedicate staff to find ways to effectively market the sample for periodic supplements while insisting that supplement costs are fully covered by the sponsors.

Conceptually, an advantage of a core-supplement design (or other design that includes a strong role for supplements) is that it allows the NCVS to play to its strengths as a *survey*.

4–C.3 Supporting Subnational Estimates

Many federal government national surveys measure key social phenomena—transportation and travel patterns or health risks, for example—that have importance to the country as a whole as well as to local areas and small subpopulations. Crime is certainly a phenomenon that has policy relevance for local, state, and federal governments. Thus, it is not surprising that a long-term tension affecting support for the NCVS has been how useful it is for local interests versus national interests.

NCVS's role as a major national social indicator (and a point of comparison with international victimization rates) notwithstanding, we think that the long-term viability of the NCVS will depend critically on its ability to provide small-domain, subnational information. "Small-domain" means subpopulations that may or may not be spatially proximate but offer important policy issues; these may include such groups as new immigrants, persons over 70 years of age living alone, or persons in areas with different growth rates (e.g., fast-growing counties, relative to stable or declining-population counties). "Subnational" refers to levels of geography smaller than the nation as a whole, such as regions, states, metropolitan statistical areas, and large cities and counties.

As Pepper and Petrie (2003:5) summarize:

> Public attention to crime and victimization often focuses on particular
> subgroups in which deviant behavior may be most troublesome. For ex-
> ample, hate crimes, crimes committed by youth, and crimes committed
> against vulnerable subpopulations including children, the elderly, and
> people with disabilities have all been the subject of recent investigations
> and legislation.

The current NCVS is capable of generating estimates by these basic demographic criteria. However, for very small subpopulations, sample sizes from
each year of the NCVS may be too small to yield stable estimates. As crime
mapping has become an increasingly useful tool in assessing trends and planning police activities, and as geographic information system (GIS) software
and Internet mapping tools have become more widespread, markets have
developed for richer spatial data on social variables like crime and victimization.

> **Recommendation 4.5:** **BJS should investigate the use of model-
> ing NCVS data to construct and disseminate subnational esti-
> mates of major crime and victimization rates.**

As described in Section 3–D.2, state and local agencies do still find value
in having a national-level measure from the NCVS as a benchmark, in many
cases because NCVS data are the only available source for some analyses.
Hence, it is an overstatement to say that the NCVS would be relevant or
important to state and local users *only* if subnational reporting was provided.
It is also the case that—perhaps with some intuitive kinds of adjustments
for differences between the national and local levels—some of the existing
demographic splits available in the NCVS can be applied to good effect.
Small states, for instance, may be able to make effective use of estimates
derived only from the NCVS sample for rural areas. When the national or
other larger area estimate is used as a proxy for a local estimate, there arises
an additional component of error because the proxy is imperfect.

A better alternative than using a proxy often can be found using methods for small-area estimation (Pfeffermann, 2002; Rao, 2003). Using such
methods, an estimate for each area of interest can be reported along with an
estimate of its mean squared error (MSE). The MSE estimates reflect both
sampling error and area-to-area variability but not biases from response errors or nonresponse errors. It is also possible to pool data over time (if
short-run change analysis is not the focus) and thereby increase the sample
size for local areas.

Small-area estimation techniques have greatly advanced since the NCVS
redesign efforts yielded the redesigned questionnaire in 1992. The NCVS

would seem to be a good candidate for evaluating small-area models of victimization estimates both for spatial domains and for demographic domains. The attraction includes the possibility of borrowing estimation strength from the police-report data of the UCR and other demographic data (perhaps enriched by the new American Community Survey data), as well as the rich covariates in NCVS, to model the victimization rates of small domains. Furthermore, the fact that the NCVS is a longitudinal study of addresses would allow the small-area models to take advantage of the covariances across time in repeated measurement areas. Since the redesign efforts, the United States has now become accustomed to small-area poverty estimates for program administration (National Research Council, 2000b,a, 1997, 1998, 1999); in terms of geography, these estimates extend to the county and school district levels and have been expanded to include estimates of health insurance programs. The Bureau of Labor Statistics Local Area Unemployment Statistics program combines data from the monthly Current Population Survey and state unemployment insurance records, along with other sources, to generate estimates for states and selected cities and counties (National Research Council, 2007:256). Folsom et al. (1999) describe methodology for producing small-area estimates of drug use, and Raghunathan et al. (2007) discuss the utilization of two survey sources to estimate cancer risk factors at the county level.

In addition to small-domain modeling using NCVS data, it may also be useful to explore ways to strengthen victimization surveys conducted by states and localities. The surveillance model (9) we describe in our table has drawbacks in its full form, as it would shift the burden for collecting victimization data on states (or groups of states) and make the "national" survey a concatenation of the state measures. Although we lean against its implementation in full, we think that the basic idea underlying the model is a sound one.

Currently, BJS operates a program under which it develops victimization survey software and provides it to interested local agencies; however, those agencies must supply all the resources (funds and manpower) to conduct a survey. An approach to strengthen this program would be to make use of BJS's organizational position within the U.S. Department of Justice. The bureau is housed in the Office of Justice Programs, the core mission of which is to provide assistance to state and local law enforcement agencies; it does so through the technical research of the National Institute of Justice and the grant programs of the Bureau of Justice Assistance (BJA), among others. We suggest that OJP consider ways of dedicating funds—like BJA grants, but separate from BJS appropriations—for helping states and localities bolster their crime information infrastructures through the establishment and regular conduct of state or regional victimization surveys.

Recommendation 4.6: BJS should develop, promote, and coordinate subnational victimization surveys through formula grants funded from state-local assistance resources.

Such surveys would most likely involve cooperative arrangements with research organizations or local universities and make use of the existing BJS statistical analysis center infrastructure.

4–C.4   Efficiency in Sample Design

The sample design of the NCVS uses data for stratification of primary and secondary units that are available from the decennial census. Stratification and the size of the sample are the two features that are tools to reduce the standard errors of NCVS estimates.

When variables are available on the frame to preidentify groups that will vary on their victimization experience, the standard errors of the estimates might be reduced. One possible approach for using external information would be to stratify areas based on other crime-related data; UCR data might be considered for this purpose.

Although the sample for the NCVS was originally designed so that each household had the same probability of being selected into the sample, such a design is not optimal for the uses of the NCVS data. Since many of the uses are at the local level, reallocating the sample to strengthen the estimates for the smaller areas will be advantageous even though the sampling variance for the largest areas (and national level) will increase, provided that those increases are not too severe. Optimization for multipurpose samples is discussed in Cochran (1977:119–123) and Kish (1976), for example. Reviewing the optimality of the numbers of primary sampling units versus number of housing units within primary sampling unit may also be beneficial.

Efficiency gains can also occur if information can be exploited to predict which housing units tend to have higher victimization rates; in that case, one can then sample the blocks containing such housing units at higher rates. Properly done, such "oversampling" can improve precision, sometimes markedly. Several approaches to predicting housing unit victimization rates may be considered:

(i) Merge block-cluster covariates into the NCVS data file and use the covariates to develop a predictive model based on past NCVS data. Prior research using the area-identified NCVS has shown that tracts with the highest levels of socioeconomic disadvantage (e.g., percentage living below poverty) have respondents with the highest victimization rates (Lauritsen, 2001).

(ii) Ask a question in the ACS about victimization and use the results to improve on the predictions in (i).

(iii) Use the first wave results for housing units to modify retention probabilities for the second wave, so that housing units that report victimization in the first wave are more likely to be retained.

Recommendation 4.7: BJS should investigate changing the sample design to increase efficiency, thus allowing more precision for a given cost. Changes to investigate include:

(i) changing the number or nature of the first-stage sampling units;

(ii) changing the stratification of the primary sampling units;

(iii) changing the stratification of housing units;

(iv) selecting housing units with unequal probabilities, so that probabilities are higher where victimization rates are higher; and

(v) alternative person-level sampling schemes (sampling or subsampling persons within housing units).

4–C.5   Other Improvements

In early redesign efforts in the early 1990s, a consistent finding that CATI interviews yielded higher reports (Hubble, 1999) led to the belief that CATI interviewing could both save money and obtain higher quality data. Now that the dispersed interviewing of the NCVS uses computer-assisted personal interviewing, the role of CATI in the overall cost-error structure of the NCVS is worth reconsidering. Both CATI and CAPI can use the same software for automatic routing and editing of responses, and the only distinction between the two approaches is connected with (a) different interviewers and (b) centralization of CATI that might affect impacts of supervision.

There are time and administrative costs from shifting cases to and from CATI facilities to field interviewers in cases in which the mode initially assigned is not desired by the respondent. These tend to reduce the response rates of the NCVS. It seems likely that the cost-quality attributes of the NCVS might be improved by shifting all telephone calling to interviewers' homes.

Recommendation 4.8: BJS should investigate the introduction of mixed mode data collection designs (including self-administered modes) into the NCVS.

As with most federal household surveys, the person-level response rates (known as Type Z rates in the NCVS) are declining over time. Whether these declines have produced estimates with higher nonresponse bias is not clear from the data at hand. Increasing evidence from survey methodology has shown that the relationship between nonresponse rates and nonresponse bias is more complicated than previously believed to be true (Curtin et al., 2000; Keeter et al., 2000; Groves, 2006).

It is quite likely that response rates will continue to fall in the NCVS. If BJS attempts to keep response rates constant, it is likely that costs of the NCVS will increase. Thus, great importance should be given to determining whether low propensities to respond to the NCVS are related to different likelihoods of different types of victimizations. There are diverse methods of such nonresponse bias studies: studying the movement of victimization rates by increasing effort to measure the cases, follow-ups of samples of nonrespondents, examining changes in estimates from alternative postsurvey adjustments, and so on. These methods and others have been noted and promoted in recent U.S. Office of Management and Budget (2006b,a) guidelines for federal surveys.

> *Recommendation 4.9:* **The falling response rates of NCVS are likely to continue, with attendant increasing field costs to avoid their decline. BJS should sponsor nonresponse bias studies, following current OMB guidelines, to guide trade-off decisions among costs, response rates, and nonresponse error.**

– 5 –

Decision-Making Process for a New Victimization Measurement System

I N THIS CHAPTER, WE FOCUS on two broader issues related to moving forward with refinements to the National Crime Victimization Survey (NCVS). The first is the need to consider ways to best develop the survey in order to shore up and expand constituencies for it (Section 5–A), and the second is the choice of the data collection agent for the survey (5–B). Several of the topics and recommendations in this chapter differ from the rest of the report in that they are agency-level in focus, aimed at better equipping the Bureau of Justice Statistics (BJS) to understand its own products and to interact with its users. This is in keeping with the panel's charge to focus on the complete portfolio of BJS programs. We make these recommendations here, in initial form, because they are pertinent to the NCVS; however, we emphasize that we expect to expand on them in our final report.

5–A BOLSTERING QUALITY AND BUILDING CONSTITUENCIES

NCVS data and estimates are routinely used by researchers and the public to understand the patterns and consequences of victimization. Researchers can access the raw data through the National Archive of Criminal Justice Data at the Interuniversity Consortium for Political and Social Research and thus can analyze the data to fit the needs of their investigation. The vast majority of the public, in contrast, has access to the data primarily through the form of routine annual estimates available on the BJS website, or through

117

special topic reports developed and released periodically on the website. However, when the public has interest in specific topics for which no regular NCVS report exists (for example, trends in rural victimization[1]), it is often beyond people's expertise to use the survey data or even to determine whether they can compile this information themselves. This problem can be addressed by using an advisory committee charged with providing BJS with information about public interest in specific kinds of NCVS reports; improving the organization of the victimization component of the BJS website so that it is clear what NCVS reports are available and what requires special analyses; and expanding the number of trend charts and spreadsheets to include compilations of interest to the public.

Any federal statistical agency must constantly strive to maintain clear communications with its users and with the best technical minds in the country relative to its data. While BJS some years ago took the initiative to stimulate the creation of the American Statistical Association's (ASA) Committee on Law and Justice Statistics, the committee is not a formal advisory committee to BJS. This means that the meetings are not public, the recommendations of the committee have no real formal documentation, and the agency does not consistently turn to the committee for key problems facing it. Furthermore, the committee consists exclusively of ASA members, who may or may not have all the expertise needed to advise BJS. A formal advisory committee has both the benefits and costs of Federal Advisory Committee Act oversight, yet it would address many of the issues cited above. Most other federal statistical agencies actively use their advisory committees (e.g., the National Center for Health Statistics, the Census Bureau, the Bureau of Labor Statistics) to seek technical input into critical challenges. This is especially true now because of the growing pressures on survey budgets arising from declining U.S. response rates.

A formal advisory committee should have membership that is appointed for its expertise. It should have experts in criminology, law enforcement, judicial processes, and incarceration. It should include state and local area experts. This expertise in the substance of the statistics should be supplemented with expertise in the methods of designing, collecting, and analyzing statistical data.

Recommendation 5.1: **BJS should establish a scientific advisory board for the agency's programs; a particular focus should be on maintaining and enhancing the utility of the NCVS.**

---

[1]Comparison of trends in urban, suburban, and rural victimization were the focus of a BJS report issued in 2000 (Duhart, 2000), but this specific analysis has not been replicated since that time.

The NCVS is largely designed and conducted for BJS by the Census Bureau. Complex survey contracts cannot be wisely administered without highly sophisticated statistical and methodological expertise. Federal statistical agencies that successfully contract out their data collection (either to the Census Bureau or a private contractor) generally have mathematical statisticians and survey methodologists who direct, coordinate, and oversee the activities of the contractor. While many of the BJS staff are labeled "statisticians," the panel observed the lack of statistical expertise that is crucial in dealing with the trade-offs of costs, sample size, numbers of primary sampling units, interviewer training, questionnaire length, use of bounding interviews, etc. The expressions of displeasure about the Census Bureau's management of the NCVS were not matched with BJS statistical analyses and simulations of design alternatives that might offer better outcomes for the agency. Furthermore, the panel thinks that the number of of BJS full-time staff dedicated to the analysis of NCVS data and the generation of reports is insufficient to exploit the full value of the survey and to navigate its challenging future. Some of the issues that require analysis (e.g., the effects of declining response rates on estimates, trade-offs of waves and questionnaire length) need statistical and methodological expertise that goes beyond current in-house capabilities.

Following the lead of other federal statistical agencies, BJS could usefully enhance statistical expertise on its staff with a program of outside research funds. When federal agencies form useful partnerships with academic researchers, they can reduce their overall costs of innovation. BJS has a track record of small research grants connected to the NCVS. The panel applauds these and urges an expansion to tackle the real methodological issues facing the NCVS.

> *Recommendation 5.2:* BJS should perform additional and advanced analysis of NCVS data. To do so, BJS should expand its capacity in the number and training of personnel and the ability to let contracts.

One reason that the panel thinks that technical staffing and external research are important is that many of the questions posed about the NCVS have not been evaluated sufficiently for us to provide recommendations to BJS on the final design of the survey. The panel thinks that this is the long-term result of "eating its seed corn," of using the operating budget too much to release the traditional reports and too little to scope out the problems of the future. It was well known 15 years ago that houschold survey response rates were falling; the impact on survey costs of these falling rates was clear (de Leeuw and de Heer, 2002). Federal statistical agencies (see CNSTAT's *Principles and Practices of a Federal Statistical Agency*) must consistently probe and analyze their own data, beyond the level required for descriptive reports, in order to see their weaknesses and their strengths. Only with such detailed knowledge can wise decisions about cost and error trade-offs be made.

> **Recommendation 5.3:** BJS should undertake research to continuously evaluate and improve the quality of NCVS estimates.

Another way that federal statistical agencies improve their data series is by nurturing a wide community of secondary analysts, using as much data as can be released within confidentiality constraints. Such analysts form a ready-made informed constituency for improving data products over time. Such analysts act as a multiplier of the impact of federal data series. Using the Internet, some agencies have expanded their impact by making available various "predigested" forms of survey data in tables, spreadsheets, graphing capabilities, etc. The panel thinks that the BJS should consider such capabilities linked to the NCVS website. These might be time series of individual population rates and means in spreadsheet form, attractive to a very broad audience, as well as microdata predesigned to have commonly desired analytic variables on observation units that are popular.

> **Recommendation 5.4:** BJS should continue to improve the availability of NCVS data and estimates in ways that facilitate user access.

BJS and the Census Bureau must keep their pledges of confidentiality to NCVS respondents. They also have the obligation to maximize the good statistical uses of the data collected with taxpayer money. Geographically identified NCVS data were available to qualified researchers from approximately 1998–2002 at the Census Bureau's research data centers (Wiersema, 1999); however, access was subsequently suspended because the data did not conform to technical conditions for research access and oversight. A project to reestablish the availability of these data by documenting and formatting internal Census Bureau data files so that they conform to Census Bureau standards began in 2005 and should be completed by the time of this report. As soon as such work is completed, these data should be made available to qualified researchers. Access to geographically identified NCVS data would permit analyses of how local characteristics and policies are associated with victimization risk and its consequences.

> **Recommendation 5.5:** The Census Bureau and BJS should ensure that geographically identified NCVS data are available to qualified researchers through the Census Bureau's research data centers, in a manner that ensures proper privacy protection.

At this writing, the U.S. statistical budget has been relatively flat for some years (except for the advent of the American Community Survey budget). These flat-line budgets have occurred at the same time that the difficulty and costs of measuring U.S. society have increased. In a climate of tight budgets and increasing costs of demographic measurement, federal statistical agencies face real threats. Such are the times that need real statistical leadership and careful stewardship of the statistical information infrastructure of the country. We fear that many surveys, the NCVS among them, can easily die "deaths from a thousand cuts." Attempts to live within the budgets lead to short-term cuts in features of surveys without certain knowledge of their effects on survey quality. Each such decision runs the risk that the country will be misled due to increased errors in data products. At some point, the basic goals of a survey cannot be met under restricted funding. The country deserves to know this when it is occurring.

The panel thinks that one opportunity for such communication comes in the annual report on statistical program funding that the U.S. Office of Management and Budget is required to prepare by a provision of the Paperwork Reduction Act of 1995 (44 U.S.C. 3504(e)(2)). This annual report—*Statistical Programs of the United States Government*—has been published for each fiscal year since 1997. The report can serve as a vehicle for alerting the executive and legislative branches to how the budget has affected the quality of statistical programs, both to the good and to the bad. With specific regard to BJS, the annual reports have generally documented the agency's responses to declining budgets. For instance, the reports for fiscal years 2007 and 2008 bore a similar warning (U.S. Office of Management and Budget, 2006c:8):

> BJS did not receive the funding requested to restore its base funding necessary to meet the growing costs of data collection and the information demands of policymakers and the criminal justice community. To address base adjustments insufficient to carry out ongoing operations of its National Crime Victimization Survey (NCVS) and other national collection programs, BJS has utilized many strategies, such as cutting sample, to keep costs within available spending levels. However, changes to the NCVS have had significant effects on the precision of the estimates—year-to-year change estimates are no longer feasible and have been replaced with two-year rolling averages.

The guidance provided by these annual reports could be enhanced through fuller explication of the impact of budget reductions (or increases) on the precision of estimates, as well as articulation of constraints and effects on federal statistical surveys systemwide. An example of the latter is the Census Bureau's sample redesign process; following the decennial census, the Census Bureau realigns the sample frames for the various demographic surveys that it conducts (including the NCVS) so that the household samples are updated and coordinated across the various data collection programs. This work is done in collaboration with the agencies that sponsor Census Bureau–conducted surveys; "the portion of the sample redesign work that can be linked to a specific survey is funded by the sponsoring agency as part of the reimbursable cost of the survey," while portions that are not directly identified with a specific survey are funded by the Census Bureau. "Thus, the approach combines central funding with user fees for survey specific redesign activities" (U.S. Office of Management and Budget, 2000:45–46). Although the sample redesign process has been routinely mentioned as an ongoing, cross-cutting activity in *Statistical Programs of the United States Government*, little detail on the progress (and consequences) of the effort was provided in the annual reports from 2001 to 2007. Ultimately, conversion from a sample deriving from the 1990 census to one using the 2000 numbers was not fully achieved for the NCVS until 2007; the redesign work was originally planned to be complete in fiscal year 2004.[2] We recommend that the annual report provide additional discussion—and warning—of budget-related effects on basic survey maintenance when appropriate.

> ***Recommendation 5.6:*** **The Statistical Policy Office of the U.S. Office of Management and Budget is uniquely positioned to identify instances in which statistical agencies have been unable to perform basic sample or survey maintenance functions. For example, BJS was unable to update the NCVS household sample to reflect population and household shifts identified in the 2000 census until 2007. The Statistical Policy Office should note such breakdowns in basic survey maintenance functions in its annual report *Statistical Programs of the United States Government*.**

5–B   DATA COLLECTION AGENT FOR THE NCVS

A review of any survey, particularly one conducted with an eye toward reducing costs, must inevitably consider the question of *who* collects the data (in addition to exactly *how* the data are collected). In the case of the NCVS, the U.S. Census Bureau of the U.S. Department of Commerce has been engaged as data collection agent since the survey's inception. In fact, as described in Box 1-1, the Census Bureau was heavily involved in the prehistory of the survey, entering into discussions with BJS's predecessor in the

---

[2]The new sample was phased in panel by panel. One panel of addresses based on the 2000 census was introduced in January 2005 for areas already included in the sample. "Beginning in January 2006, [the Census Bureau] introduced sample based on the 2000 decennial census in new areas. The phase-in of the 2000 sample and the phase-out of the 1990 sample will be complete in January 2008" (Demographic Surveys Division, U.S. Census Bureau, 2007b).

late 1960s and convening planning conferences that would give shape to the NCVS and its pretests. Since "it was clear from the pilot studies that large samples would be required to obtain reliable estimates of victimization for crime classes of intense interest (e.g., rape)," "the Census Bureau was the only organization that could field such a large survey" and hence was the natural choice as the data collection agent for the new NCVS (Cantor and Lynch, 2000:105).

The choice of the Census Bureau as the data collector for the NCVS had implications for the survey's design, as summarized by Cantor and Lynch (2000:107):

> Other design features of NCS were occasioned by the need to fit into the organization of the Census Bureau and the Current Population Survey (CPS). CPS is the largest intercensal survey conducted in the world and, at the time, NCS was to be the second largest of these surveys. Sharing interviewers between the two surveys would mean great efficiencies for the [Census Bureau]. CPS employed a rotating panel design. This was viewed as an advantage to NCS for a number of reasons. One was the ability to use prior interviews to 'bound' subsequent interviews. . . . A second was that the rotating panel design substantially increased the precision of the year-to-year change estimates. The panel design feature produces a natural positive correlation across annual estimates. This, in turn, substantially reduces the standard error on change estimates.

As may be expected, the experience of decades of work has illustrated both advantages and disadvantages of the relationship between BJS as sponsor and funder of the NCVS and the Census Bureau as its data collector. Relatively few of the conceptual pros and cons are unique to the BJS-Census relationship; rather, they are generally applicable to any contractor and client.

Others, however, in the panel's view deserve comment. A basic concern that has arisen about the Census Bureau as the data collection agent for the NCVS is the lack of transparency in costs. Historically, the Census Bureau has not provided its federal agency survey sponsors with detailed breakdowns in survey costs (and rationales for changes in costs, over and above the known increasing costs of gaining compliance in survey research). It is the panel's view that disaggregated costs are key to effective innovation in large-scale surveys. The data collector must know what survey design choices are associated with the largest portions of costs in order to effectively consider trade-offs of costs and errors. Recent attention to survey costs (e.g., at conferences hosted by the Federal Committee on Statistical Methodology and the National Institute of Statistical Sciences) have shown the value of detailed cost accounting.[3]

---

[3] See http://www.fcsm.gov/events/program/2006FCSMFinalprogram.pdf (see the session on "modeling survey costs"); Karr and Last (2006).

Recommendation 5.7: Because BJS is currently receiving inadequate information about the costs of the NCVS, the Census Bureau should establish a data-based, data-driven survey cost and information system.

Some of the features of the NCVS are not shared by other designs and, lacking a strong evidentiary base for their choice, this stimulates the panel to wonder why the Census Bureau and BJS have chosen them. These include the recycling of cases from the field to centralized computer-assisted telephone interviewing (CATI) (instead of using a dispersed field interviewing corps for the telephone interviews). They include the slowness of moving from paper questionnaires to computer-assisted personal interviewing (CAPI). They include the failure to study the use of audio computer-assisted interviewing for many of the sensitive topics in the survey, despite its widespread use in other federal surveys (e.g., the National Survey of Drug Use and Health and the National Survey of Family Growth, as well as BJS-sponsored data collections as required by the Prison Rape Elimination Act). They include the lack of study of how best to use the bounding interview in estimation. Finally, the panel notes that there is very little substantive expertise in criminology and justice programs within the Census Bureau staff working on the NCVS. That means that the Census Bureau focuses on field and statistical issues without the advantage of formal educational background in the substance of the NCVS. Just as the BJS staff would be stronger with more technical and statistical expertise, the panel thinks that the Census Bureau could mount a better NCVS and partner more effectively with BJS with more substantive expertise.

That said, it must be noted with equal force that there are important advantages to the use of the Census Bureau as data collector. Census Bureau household surveys, by and large, achieve higher response rates than comparable surveys conducted by a private contractor on behalf of the federal government. It is common throughout the world that central government statistical agencies achieve higher response rates than private-sector survey organization (Groves and Couper, 1998). The Census Bureau has maintained a strong confidentiality pledge through the force of the Title 13 law, although under the widened protection of the Confidential Information Protection and Statistical Efficiency Act of 2002, it is not clear that that advantage will be maintained. Furthermore, interagency agreements within the federal government appear to be simpler and less burdened by regulation than federal contracts. Finally—in the event that a radical option for collecting victimization data were necessary—continued partnership with the Census Bureau could offer the benefit of more readily piggybacking some victimization measures on one of the Census Bureau's ongoing surveys (e.g., the American Community Survey or Current Population Survey; see Section 4–B.1).

BJS has sought input regarding contracting out the NCVS to the private sector. We urge careful consideration of survey cost structures prior to such a move. The panel notes that this review would be greatly facilitated if BJS could obtain disaggregated costs from the Census Bureau for the current NCVS. BJS should study other federal surveys contracted out to the private sector to determine the extent to which flexibility in dealing with changes and innovations was or was not realized. It should also study the implications of contracting out on the desired staff skills within BJS.

One way to increase understanding of the trade-offs of different NCVS designs and different contracting models is to seek formal design alternatives from the Census Bureau and others. A formal design competition could be mounted, perhaps through a set of commissioned designs, both from the Census Bureau and other survey methodologists. The designs would be guided by the same goals, articulated by BJS, but would be left to the creativity of the designers. The design options should be costed out in as much detail as possible, and the designs should be critiqued through peer review.

Recommendation 5.8: **BJS should consider a survey design competition in order to get a more accurate reading of the feasibility of alternative NCVS redesigns. The design competition should be administered with the assistance of external experts, and the competition should include private organizations under contract and the Census Bureau under an interagency agreement.**

References

Addington, L. A. (2005). Disentangling the effects of bounding and mobility on reports of criminal victimization. *Journal of Quantitative Criminology 21*(3), 321–343.

Addington, L. A. (2007). Using NIBRS to study methodological sources of divergence between the UCR and NCVS. See Lynch and Addington (2007), Chapter 8.

Alaska Justice Statistical Analysis Center (2002, February). *Measuring Adult Victimization in Alaska: Technical Report*. Report for the Bureau of Justice Statistics. JC 20 | 0109.011. Anchorage: Justice Center, University of Alaska Anchorage.

Anderson, D. A. (1999). The aggregate burden of crime. *Journal of Law and Economics 42*(611).

Atrostic, B., N. Bates, G. Burt, and A. Silberstein (2001). Nonresponse in U.S. government household surveys: Consistent measures, recent trends, and new insights. *Journal of Official Statistics 17*(2), 209–226.

Australian Bureau of Statistics (2006, April). *Crime and Safety, Australia—2005*. Release 4509.0. Canberra: Australian Bureau of Statistics.

Bachman, R., H. Dillaway, and M. S. Lachs (1998). Violence against the elderly: A comparative analysis of robbery and assault across age and gender groups. *Research on Aging 20*(2), 183–198.

Bachman, R., L. E. Saltzman, M. P. Thompson, and D. C. Carmody (2002). Disentangling the effects of self-protective behaviors on the risk of injury in assaults against women. *Journal of Quantitative Criminology 18*(2), 135–157.

Bachman, R. and B. M. Taylor (1994). The measurement of family violence and rape by the redesigned National Crime Victimization Survey. *Justice Quarterly 11*, 499–512.

Bates, N. (2006, December 29). Noninterview rates for selected major demographic household surveys, 1990–2005. Internal memorandum, Demographic Surveys Division, U.S. Census Bureau.

Baum, K. (2007, November). *National Crime Victimization Survey: Identity Theft, 2005.* NCJ 219411. Washington, DC: Bureau of Justice Statistics.

Baum, K. and P. Klaus (2005, January). *National Crime Victimization Survey: Violent Victimization of College Students, 1995–2002.* NCJ 206836. Washington, DC: Bureau of Justice Statistics.

Baumer, E. P. (2002). Neighborhood disadvantage and police notification by victims of violence. *Criminology 40*(3), 579–616.

Baumer, E. P., J. Horney, R. B. Felson, and J. L. Lauritsen (2003). Neighborhood disadvantage and the nature of violence. *Criminology 41*(1), 39–71.

Belli, R. F. (1998). The structure of autobiographical memory and the event history calendar: Potential improvements in the quality of retrospective reports in surveys. *Memory 6*(4), 383–406.

Belli, R. F., W. L. Shay, and F. P. Stafford (2001). Event history calendars and question list surveys: A direct comparison of interviewing methods. *Public Opinion Quarterly 65*, 45–74.

Belli, R. F., L. M. Smith, P. M. Andreski, and S. Agrawal (2007). Methodological comparisons between CATI event history calendar and standardized conventional questionnaire instruments. *Public Opinion Quarterly 71*(4), 603–622.

Bensley, L., M. S., J. Van Eenwyk, K. Wynkoop Simmons, and D. Ruggles (2000, July 7). Prevalence of intimate partner violence and injuries—Washington, 1998. *Morbidity and Mortality Weekly Reports 49*(26), 589–592.

Biderman, A. D. and D. Cantor (1984). A longitudinal analysis of bounding respondent conditioning and mobility as sources of panel bias in the National Crime Survey. In *Proceedings of the Survey Methods Research Section.* Alexandria, VA: American Statistical Association.

Biderman, A. D., D. Cantor, J. P. Lynch, and E. Martin (1986). *Final Report of Research and Development for the Redesign of the National Crime Survey.* Prepared for the Bureau of Justice Statistics. Washington, DC: Bureau of Social Science Research, Inc.

Biderman, A. D., D. Cantor, and A. Reiss (1985). A quasi-experimental analysis of personal victimization by household respondents in the NCS. Paper presented at the annual meetings of the American Statistical Association, Philadelphia.

Biderman, A. D. and A. J. Reiss (1967, November). On exploring the "dark figure" of crime. *Annals of the American Academy of Political and Social Science 374*, 1–15.

Bijleveld, C. C. and P. R. Smit (2004, September). Netherlands. See Farrington et al. (2004), Chapter 6, pp. 151–186.

Bradburn, N. M., J. Huttenlocher, L. Hedges, N. Schwartz, and S. Sudman (1994). Telescoping and temporal memory. In N. Schwartz and S. Sudman (Eds.), *Autobiographical Memory and the Validity of Retrospective Reports*. New York: Springer-Verlag.

Bureau of Justice Statistics (1989). *Redesign of the National Crime Survey*. NCJ 111457. Washington, DC: U.S. Department of Justice.

Bureau of Justice Statistics (2000, May). *Criminal Victimization in the United States, 1995: A National Crime Victimization Survey Report*. NCJ 171129. Washington, DC: U.S. Department of Justice.

Bureau of Justice Statistics (2006a, December). *Criminal Victimization in the United States, 2005—Statistical Tables*. NCJ 215244. Washington, DC: U.S. Department of Justice.

Bureau of Justice Statistics (2006b). *National Crime Victimization Survey Longitudinal File, 1995–1999*. Computer file. Conducted by U.S. Department of Commerce, Bureau of the Census. ICPSR 04414. Ann Arbor, MI: Inter-university Consortium for Political and Social Research [producer and distributor].

Bureau of Justice Statistics (2006c). *State Justice Statistics Fiscal Year 2006 Program Announcement*. Washington, DC: U.S. Department of Justice, Office of Justice Programs.

Bureau of Justice Statistics (2007a). *National Crime Victimization Survey, 2005*. Computer file. Conducted by U.S. Department of Commerce, Bureau of the Census. ICPSR 04451. Ann Arbor, MI: Inter-university Consortium for Political and Social Research [producer and distributor].

Bureau of Justice Statistics (2007b, January 15). *National Crime Victimization Survey: MSA Data, 1979–2004*. Computer file. Conducted by U.S. Department of Commerce, Bureau of the Census. ICPSR 04576. Ann Arbor, MI: Inter-university Consortium for Political and Social Research [producer and distributor].

Bureau of Labor Statistics (2007, February 16). *2005 Consumer Expenditure Diary Survey, Public Use Microdata Documentation*. Washington, DC: U.S. Department of Labor, Bureau of Labor Statistics, Division of Consumer Expenditure Surveys.

Bushery, J. M. (1981a). Recall biases for different reference periods in the National Crime Survey. In *Proceedings of the Survey Methods Research Section*, pp. 238–243. Alexandria, VA: American Statistical Association.

Bushery, J. M. (1981b, March 31). *Results of the NCS Reference Period Research Experiment*. Memorandum for the NCS Reference Committee. Washington, DC: U.S. Bureau of the Census.

Cantor, D. and J. P. Lynch (2000). Self-report surveys as measures of crime and criminal victimization. In *Measurement and Analysis of Crime and Justice*, Volume 4 of *Criminal Justice 2000*, pp. 85–138. NCJ 182411. Washington, DC: U.S. Department of Justice.

Cantor, D. and J. P. Lynch (2005). Exploring the effects of changes in design on the analytical uses of the NCVS data. *Journal of Quantitative Criminology 21*(3), 293–319.

Carcase, C. (2004, September). Australia. See Farrington et al. (2004), Chapter 4, pp. 75–118.

Catalano, S. M. (2006, September). *National Crime Victimization Survey: Criminal Victmization, 2005*. NCJ 214644. Washington, DC: U.S. Department of Justice.

Catalano, S. M. (2007). Methodological change in the NCVS and the effect on convergence. See Lynch and Addington (2007), Chapter 5.

Centers for Disease Control and Prevention (2003). Public health surveillance for behavioral risk factors in a changing environment: Recommendations from the Behavioral Risk Factor Surveillance Team. *Morbidity and Mortality Weekly Report 52*(RR-9), 1–12.

Centers for Disease Control and Prevention (2005, September 2). *Behavioral Risk Factor Surveillance System Questionnaire*. Behavioral Surveillance Branch. Atlanta: U.S. Department of Health and Human Services, Centers for Disease Control and Prevention.

Centers for Disease Control and Prevention (2006a). *Behavioral Risk Factor Surveillance System Operational and User's Guide*. Version 3.0. Behavioral Surveillance Branch. Atlanta: U.S. Department of Health and Human Services, Centers for Disease Control and Prevention.

Centers for Disease Control and Prevention (2006b, January 23). *Behavioral Risk Factor Surveillance System Questionnaire*. Behavioral Surveillance Branch. Atlanta: U.S. Department of Health and Human Services, Centers for Disease Control and Prevention.

Centers for Disease Control and Prevention (2007, February 14). *2006 Behavioral Risk Factor Surveillance System Data Quality Report Handbook*. Version 1.0.0. Behavioral Surveillance Branch. Atlanta: U.S. Department of Health and Human Services, Centers for Disease Control and Prevention.

Central Statistics Office (2007, April 25). *Crime and Victimisation: Quarterly National Household Survey, 2006 (including results for 1996 and 2003)*. Ref 75/2007. Dublin: Central Statistics Office.

Clements, W. and M. Bellas (2003). *Criminal Victimization in Vermont, 2001*. Montpelier, VT: Vermont Center for Justice Research.

Cochran, W. G. (1977). *Sampling Techniques* (3rd ed.). New York: Wiley.

Cohen, M. A. (1988). Pain suffering and jury awards: A study of the cost of crime to victims. *Law and Society Review 22*, 538–555.

Cohen, M. A. (2000). Measuring the costs and benefits of crime and justice. *Criminal Justice 4*, 264–315.

Cohen, M. A. (2005). *The Costs of Crime and Justice*. London and New York: Routledge.

Conaway, M. and S. Lohr (1994). A longitudinal analysis of factors associated with the reporting of violent crime to the police. *Journal of Quantitative Criminology 10*, 23–39.

Cook, P. J. (1985). The case of the missing victims: Gunshot woundings in the National Crime Survey. *Journal of Quantitative Criminology 1*(1), 91–102.

Cowan, C. D., L. R. Murphy, and J. Wiener (1984). Effects of supplemental questions on victimization estimates. See Lehnen and Skogan (1984), pp. 69–73.

Curtin, R., S. Presser, and E. Singer (2000). The effects of response rate changes on the index of consumer sentiment. *Public Opinion Quarterly 64*(4), 413–428.

Czaja, R., J. Blair, B. Bickart, and E. Eastman (1994). Respondent strategies for recall of crime victimization incidents. *Journal of Official Statistics 10*(257–276).

de Leeuw, E. and W. de Heer (2002). Trends in household survey nonresponse: A longitudinal and international comparison. In R. Groves, D. Dillman, J. Eltinge, and R. Little (Eds.), *Survey Nonresponse*, pp. 41–54. New York: Wiley.

DeFrances, C. J. and S. K. Smith (1994, June). *Crime and Neighborhoods*. NCJ 147005. Washington, DC: Bureau of Justice Statistics.

DeFrances, C. J. and S. K. Smith (1998, April). *Perceptions of Neighborhood Crime, 1995*. NCJ 165811. Washington, DC: Bureau of Justice Statistics.

DeMaio, T. J. (1980). Refusals: Who, where, and why. *Public Opinion Quarterly 44*(2), 223–233.

Demographic Surveys Division, U.S. Census Bureau (2007a, April 13). *Key Points of Interest Regarding the National Crime Victimization Survey (NCVS)*. Material provided to accompany comments to the Panel to Review the Programs of the Bureau of Justice Statistics. Washington, DC: U.S. Census Bureau.

Demographic Surveys Division, U.S. Census Bureau (2007b, January). *Survey Abstracts*. Washington, DC: U.S. Census Bureau.

Dinkes, R., E. F. Cataldi, W. Lin-Kelly, and T. D. Snyder (2007, December). *Indicators of School Crime and Safety: 2007*. NCES 2008-021; NCJ 219553. Washington, DC: U.S. Department of Education and U.S. Department of Justice, Office of Justice Programs.

Dugan, L. (1999). The effect of criminal victimization on a household's moving decision. *Criminology 37*(4), 903–930.

Dugan, L. and R. Apel (2003). An exploratory study of the violent victimization of women: Race/ethnicity and situational context. *Criminology 41*(3), 959–980.

Duhart, D. T. (2000, October). *Urban, Suburban, and Rural Victimization, 1993–98.* NCJ 182031. Washington, DC: U.S. Department of Justice, Bureau of Justice Statistics.

Durose, M. R., C. W. Harlow, P. A. Langan, M. Motivans, R. R. Rantala, and E. L. Smith (2005, June). *Family Violence Statistics: Including Statistics on Strangers and Acquaintances.* NCJ 207846. Washington, DC: U.S. Department of Justice.

Engel, R. S. (2005). Citizens' perceptions of distributive and procedural injustice during traffic stops with police. *Journal of Research in Crime and Delinquency 42*(4), 445–481.

Engel, R. S. and J. M. Calnon (2004). Examining the influence of drivers' characteristics during traffic stops with police: Results from a national survey. *Justice Quarterly 21*(1), 49–90.

Farrington, D. P., P. A. Langan, and M. Tonry (Eds.) (2004, September). *Cross-National Studies in Crime and Justice.* NCJ 200988. Washington, DC: U.S. Department of Justice.

Federal Bureau of Investigation (2004). *UCR: Uniform Crime Reporting Handbook.* Washington, DC: U.S. Department of Justice.

Federal Interagency Forum on Child and Family Statistics (2007). *America's Children: Key National Indicators of Well-Being, 2007.* Washington, DC: U.S. Government Printing Office.

Folsom, R., B. Shah, and A. Vaish (1999). Substance abuse in states: A methodological report on model based estimates from the 1994–1996 national household surveys on drug abuse. In *Proceedings of the Survey Methods Research Section*, pp. 311–375. Alexandria, VA: American Statistical Association.

Gannon, M. (2006). Crime statistics in Canada, 2005. *Juristat: Canadian Centre for Justice Statistics 26*(4), 1–23. Catalogue no. 85-002-XIE.

Giblin, M. (2003, July). *Measuring Adult Criminal Victimization: Findings from the Anchorage Adult Criminal Victimization Survey.* Report for the Alaska Justice Statistical Analysis Center to the Bureau of Justice Statistics. JC 0109.021. Anchorage: Justice Center, University of Alaska Anchorage.

Grant, C., K. Bolling, and M. Sexton (2007). *2005–6 British Crime Survey (England and Wales): Technical Report, Volume I.* Prepared for Research, Development and Statistics—Strategic Data Flows (Crime Reduction and Community Safety Group). London: Home Office, United Kingdom.

Gray, C. M. (Ed.) (1979). *The Costs of Crime.* Beverly Hills, CA: Sage Publications.

Greenfeld, L. A. and M. A. Henneberg (2001). Victim and offender self-reports of alcohol involvement in crime. *Alcohol Research and Health 25*(1).

Greenfeld, L. A., P. A. Langan, and S. K. Smith (1997). *Police Use of Force: Collection of National Data.* NCJ 165040. Washington, DC: Bureau of Justice Statistics.

Griffin, D. H., J. K. Broadwater, T. F. Leslie, P. D. McGovern, and D. A. Raglin (2004, December). *Meeting 21st Century Demographic Data Needs— Implementing the American Community Survey; Report 11: Testing Voluntary Methods—Additional Results.* Washington, DC: U.S. Census Bureau.

Griffin, D. H., D. A. Raglin, T. F. Leslie, P. D. McGovern, and J. K. Broadwater (2003, December). *Meeting 21st Century Demographic Data Needs— Implementing the American Community Survey; Report 3: Testing the Use of Voluntary Methods.* Washington, DC: U.S. Census Bureau.

Groves, R. M. (2006). Nonresponse rates and nonresponse bias in household surveys. *Public Opinion Quarterly 70*(5), 646–675.

Groves, R. M. and M. P. Couper (1998). *Nonresponse in Household Interview Surveys.* New York: Wiley.

Groves, R. M., D. A. Dillman, J. L. Eltinge, and R. J. Little (Eds.) (2002). *Survey Nonresponse.* New York: Wiley.

Groves, R. M., S. Presser, and S. Dipko (2004). The role of topic interest in survey participation decisions. *Public Opinion Quarterly 68*, 2–31.

Haddon, M. and J. Christenson (2005). *Shedding Light: 2004 Utah Crime Victmization Survey.* Salt Lake City: Utah Commission on Criminal and Juvenile Justice, Research and Data Unit.

Harlow, C. W. (2005, November). *National Crime Victimization Survey and Uniform Crime Reporting: Hate Crime Reported by Victims and Police.* NCJ 209911. Washington, DC: Bureau of Justice Statistics.

Hart, T. C. and C. M. Rennison (2003). *Reporting Crime to the Police, 1992–2000.* NCJ 195710. Washington, DC: Bureau of Justice Statistics.

Hiselman, J., P. Stevenson, G. Ramker, D. Olson, and L. Levinson (2005). *The Extent and Nature of Adult Crime Victimization in Illinois, 2002.* Chicago: Illinois Criminal Justice Information Authority.

Home Office (2007, February). *Guidance on Statutory Performance Indicators for Policing, 2006/07, Version 4.* Police and Crime Standards Directorate. London: Home Office, United Kingdom.

Hope, S. (2005). *Scottish Crime and Victimisation Survey Calibration Exercise: A Comparison of Survey Methodologies.* Edinburgh: Scottish Executive.

Hubble, D. L. (1999). NCVS: New questionnaire and procedures development and phase-in methodology. In *Proceedings of the Survey Methods Research Section*, pp. 63–72. Alexandria, VA: American Statistical Association.

Iachan, R., J. Schulman, E. Powell-Griner, D. E. Nelson, P. Mariolis, and C. Stanwyck (2001). Pooling state telephone survey health data for national estimates: The CDC Behavioral Risk Factor Surveillance System, 1995. In M. L. Cynamon and R. A. Kulka (Eds.), *Conference on Health Survey Research Methods*, Number 7th, DHHS Publication No. (PHS) 01-1013. Hyattsville, MD, pp. 221–226. Conference on Health Survey Research Methods: National Center for Health Statistics.

Jansson, K. (2007). *British Crime Survey—Measuring Crime for 25 Years*. London: Home Office, United Kingdom.

Johnson, H. (2005). *Crime Victimisation in Australia: Key Results of the 2004 International Crime Victimisation Survey*. Number 64 in Research and Public Policy. Canberra: Australian Institute of Criminology.

Karmen, A. (2007). *Crime Victims: An Introduction to Victimology* (6th ed.). Belmont, CA: Thomson Wadsworth.

Karr, A. F. and M. Last (2006, September 29). *Survey Costs: Workshop Report and White Paper*. Research Triangle Park, North Carolina: National Institute of Statistical Sciences. Summarizes a workshop held April 18–19, 2006, at the National Center for Education Statistics, Washington, DC.

Keeter, S., C. Miller, A. Kohut, R. M. Groves, and S. Presser (2000). Consequences of reducing nonresponse in a national telephone survey. *Public Opinion Quarterly 64*(2), 125–148.

Killias, M., P. Lamon, and M. F. Aebi (2004, September). Switzerland. See Farrington et al. (2004), Chapter 9, pp. 239–302.

Kindermann, C., J. Lynch, and D. Cantor (1997). *Effects of the Redesign on Victimization Estimates*. NCJ 164381. Washington, DC: U.S. Department of Justice.

Kish, L. (1976). Optima and proxima in linear sample designs. *Journal of the Royal Statistical Society, Series A 139*, 80–95.

Klaus, P. (1999, March). *Carjackings in the United States, 1992–96*. NCJ 171145. Washington, DC: Bureau of Justice Statistics.

Klaus, P. (2004, July). *National Crime Victimization Survey: Carjacking, 1993–2002*. NCJ 205123. Washington, DC: Bureau of Justice Statistics.

Klaus, P. (2007, April). *National Crime Victimization Survey: Crime and the Nation's Households, 2005*. NCJ 217198. Washington, DC: U.S. Department of Justice.

Klaus, P. and C. M. Rennison (2002, February). *Age Patterns in Violent Victimization, 1976–2000*. NCJ 190104. Washington, DC: Bureau of Justice Statistics.

Klaus, P. A. (1994). *The Costs of Crime to Victims: Crime Data Brief.* NCJ 145865. Washington, DC: U.S. Department of Justice, Bureau of Justice Statistics.

Lauritsen, J. L. (2001). The social ecology of violent victimization: Individual and contextual effects in the NCVS. *Journal of Quantitative Criminology 17*(1), 3–32.

Lauritsen, J. L. (2005). Social and scientific influences on the measurement of criminal victimization. *Journal of Quantitative Criminology 21*(3), 245–266.

Lauritsen, J. L. (2006a, August). Criminal victimization in American metropolitan areas. Paper presented at the annual meeting of the American Sociological Association, Montreal, Canada.

Lauritsen, J. L. (2006b, February 17). *Evaluation of NCVS Methodology and Costs: Report for the Bureau of Justice Statistics.* Final Report for Visiting Research Fellowship, Grant (2005-BJ-CX-K025). Washington, DC: U.S. Department of Justice.

Lauritsen, J. L. and R. Schaum (2004). The social ecology of violence against women. *Criminology 42*(2), 323–358.

Lauritsen, J. L. and R. J. Schaum (2005). *Crime and Victimization in the Three Largest Metropolitan Areas, 1980–98.* NCJ 208075. Washington, DC: U.S. Department of Justice.

Law Enforcement Assistance Administration (1981). Current objectives for the National Crime Survey program. See Lehnen and Skogan (1981), pp. 10. Excerpted from a memorandum from James Gregg, Acting Administrator, Law Enforcement Assistance Administration, to Peter F. Flaherty, Deputy Attorney General, entered as testimony before the Committee on the Judiciary, U.S. House of Representatives, October 13, 1977.

Lee, M., D. Lewis, M. Crowley, E. Hock, C. Laskey, C. Loftin, W. Logan, and L. Addington (1999). Developing hate crime questions for the National Crime Victimization Survey. In *Proceedings of the Survey Methods Research Section*, pp. 1036–1041. Alexandria, VA: American Statistical Association.

Lehnen, R. G. and W. G. Skogan (Eds.) (1981). *The National Crime Survey Working Papers: Volume I: Current and Historical Perspectives.* Number NCJ 75374. Washington, DC: U.S. Department of Justice.

Lehnen, R. G. and W. G. Skogan (Eds.) (1984). *The National Crime Survey Working Papers: Volume II: Methodological Studies.* Number NCJ 90307. Washington, DC: U.S. Department of Justice.

Lepkowski, J. P. and M. P. Couper (2002). Nonresponse in longitudinal household surveys. See Groves et al. (2002), Chapter 17, pp. 259–272.

Levitt, S. D. (1999). The changing relationship between income and crime victimization. *Economic Policy Review 5*(3), 993–1016.

Lewis, D. (2002, June). NCVS research. NCVS Research Unit, Demographic Surveys Division, U.S. Census Bureau, Memorandum 02-02.

Lynch, J. P. (2002). *Trends in Juvenile Violent Offending.* Juvenile Justice Bulletin. NCJ 191052. Washington, DC: U.S. Department of Justice, Office of Juvenile Justice and Delinquency Prevention.

Lynch, J. P. (2006). Problems and promise of victimization surveys for cross-national research. *Crime and Justice: A Review of Research 34.* Accessed through Lexis-Nexis; pagination differs from printed volume.

Lynch, J. P. and L. A. Addington (Eds.) (2007). *Understanding Crime Statistics: Revisiting the Divergence of the NCVS and UCR.* Cambridge Studies in Criminology. New York: Cambridge University Press.

Lynch, J. P., M. L. Berbaum, and M. Planty (1998). *Investigating Repeated Victimization with the NCVS.* Final report for National Institute of Justice Grant 97-IJ-CX-0027. Washington, DC: U.S. Department of Justice.

Lynn, P. (1997). *Collecting Data About Non-Respondents to the British Crime Survey.* London: Social and Community Planning Research.

Maltz, M. D. (1999, September). *Bridging Gaps in Police Crime Data: A Discussion Paper from the BJS Fellows Program.* Bureau of Justice Statistics and Federal Bureau of Investigation Uniform Crime Reporting Program. NCJ 176365. Washington, DC: U.S. Department of Justice.

Maltz, M. D. (2007). Missing UCR data and divergence of the NCVS and UCR trends. See Lynch and Addington (2007), Chapter 10.

Martin, E., R. M. Groves, J. Matlin, and C. Miller (1986). *Report on the Development of Alternative Screening Procedures for the National Crime Survey.* Washington, DC: Bureau of Social Science Research, Inc.

May, D., J. Wells, K. Minor, K. Cobb, E. Angel, and K. Cline (2004). *Crime Victimization Experiences, Fear of Crime, Perceptions of Risk, and Opinion of Criminal Justice Agents Among a Sample of Kentucky Residents.* Center for Criminal Justice Education Research, Department of Correctional and Juvenile Justice Studies, College of Justice and Safety. Richmond, KY: Eastern Kentucky University.

McDowall, D. and C. Loftin (2007). What is convergence, and what do we know about it? See Lynch and Addington (2007), Chapter 4.

McManus, R. (2002). *Criminal Victimization Trends in South Carolina.* Blythewood, SC: Office of Justice Programs, South Carolina Department of Public Safety.

Miller, P. V. and R. M. Groves (1985). Matching survey responses to official records: An exploration of validity in victimization reporting. *The Public Opinion Quarterly 49*(3), 366–380.

Miller, T. R., M. A. Cohen, and B. Wiersema (1996). *Victim Costs and Consequences: A New Look.* Research Report. NCJ 155282. Washington, DC: U.S. Department of Justice, National Institute of Justice.

Minnesota Justice Statistics Center (2003). *Safe at Home: 2002 Minnesota Crime Survey.* Minnesota Planning. St. Paul, MN: Minnesota Department of Public Safety, Office of Justice Programs.

Murphy, L. (1976). *The Effects of the Attitude Supplement on NCS City Sample Victimization Data.* Unpublished internal document. Washington, DC: U.S. Bureau of the Census.

Murphy, L. R. (1984). Effects of bounding on telescoping in the National Crime Survey. See Lehnen and Skogan (1984), pp. 83–89.

National Center for Victims of Crime (2003). *Extending Our Reach, Reaching More: Victims of Crime, 2000–2002 Progress Report.* Washington, DC: National Center for Victims of Crime.

National Research Council (1976a). *Setting Statistical Priorities.* Washington, DC: National Academy Press.

National Research Council (1976b). *Surveying Crime.* Panel for the Evaluation of Crime Surveys. Bettye K. Eidson Penick, editor, and Maurice E.B. Owens III, associate editor. Committee on National Statistics, Academy of Mathematical and Physical Sciences. Washington, DC: National Academy of Sciences.

National Research Council (1984). *Cognitive Aspects of Survey Methodology: Building a Bridge Between Disciplines.* Report of the Advanced Research Seminar on Cognitive Aspects of Survey Methodology. Thomas B. Jabine, Miron L. Straf, Judith M. Tanur, and Roger Tourangeau (Eds.), Committee on National Statistics, Commission on Behavioral and Social Sciences and Education. Washington, DC: National Academy Press.

National Research Council (1997). *Small-Area Estimates of School-Age Children in Poverty: Interim Report 1, Evaluation of 1993 County Estimates for Title I Allocations.* Panel on Estimates of Poverty for Small Geographic Areas, Constance F. Citro, Michael L. Cohen, Graham Kalton, and Kirsten K. West (Eds.), Committee on National Statistics. Washington, DC: National Academy Press.

National Research Council (1998). *Small-Area Estimates of School-Age Children in Poverty: Interim Report 2, Evaluation of Revised 1993 County Estimates for Title I Allocations.* Panel on Estimates of Poverty for Small Geographic Areas, Constance F. Citro, Michael L. Cohen, and Graham Kalton (Eds.), Committee on National Statistics. Washington, DC: National Academy Press.

National Research Council (1999). *Small-Area Estimates of School-Age Children in Poverty: Interim Report 3.* Panel on Estimates of Poverty for Small Geographic Areas, Constance F. Citro and Graham Kalton (Eds.), Committee on National Statistics. Washington, DC: National Academy Press.

National Research Council (2000a). *Small-Area Estimates of School-Age Children in Poverty: Evaluation of Current Methodology.* Panel on Estimates of Poverty for Small Geographic Areas, Constance F. Citro and Graham Kalton (Eds.), Committee on National Statistics. Washington, DC: National Academy Press.

National Research Council (2000b). *Small-Area Income and Poverty Estimates: Priorities for 2000 and Beyond.* Panel on Estimates of Poverty for Small Geographic Areas, Constance F. Citro and Graham Kalton (Eds.), Committee on National Statistics. Washington, DC: National Academy Press.

National Research Council (2003a). *Measurement Problems in Criminal Justice Research: Workshop Summary.* Committee on Law and Justice and Committee on National Statistics, John V. Pepper and Carol V. Petrie (Eds.), Division of Behavioral and Social Sciences and Education. Washington, DC: The National Academies Press.

National Research Council (2003b). *Survey Automation: Report and Workshop Proceedings.* Oversight Committee for the Workshop on Survey Automation, Daniel L. Cork, Michael L. Cohen, Robert Groves, and William Kalsbeek (Eds.), Committee on National Statistics. Washington, DC: The National Academies Press.

National Research Council (2004). *Principles and Practices for a Federal Statistical Agency* (Third ed.). Committee on National Statistics. Margaret E. Martin, Miron L. Straf, and Constance F. Citro (Eds.), Division of Behavioral and Social Sciences and Education. Washington, DC: The National Academies Press.

National Research Council (2006). *Once, Only Once, and in the Right Place: Residence Rules in the Decennial Census.* Panel on Residence Rules in the Decennial Census. Daniel L. Cork and Paul R. Voss (Eds.), Committee on National Statistics, Division of Behavioral and Social Sciences and Education. Washington, DC: The National Academies Press.

National Research Council (2007). *Using the American Community Survey: Benefits and Challenges.* Panel on the Functionality and Usability of Data from the American Community Survey. Constance F. Citro and Graham Kalton (Eds.), Committee on National Statistics, Division of Behavioral and Social Sciences and Education. Washington, DC: The National Academies Press.

O'Brien, R. (1996). Police productivity and crime rates: 1973–1992. *Criminology 34*, 183–207.

Pepper, J. V. and C. V. Petrie (2003). Introduction. See National Research Council (2003a), Chapter 1, pp. 1–9.

Pfeffermann, D. (2002). Small area estimation—new developments and directions. *International Statistical Review 70*(1), 125–143.

Planty, M. (2007). Series victimization and divergence. See Lynch and Addington (2007), Chapter 6.

Planty, M. and K. Strom (2007). Understanding the role of repeat victims in the production of annual US victimization rates. *Journal of Quantitative Criminology 23*(3), 179–200.

President's Commission on Law Enforcement and Administration of Justice (1967, February). *The Challenge of Crime in a Free Society.* Washington, DC: U.S. Government Printing Office. Commonly known as the Crime Commission report.

Raghunathan, T. E., D. Zie, N. Schenker, V. L. Parsons, W. W. Davis, K. W. Dodd, and E. J. Feuer (2007). Combining information from two surveys to estimate county-level prevalence rates of cancer risk factors and screening. *Journal of the American Statistical Association 102*(478), 474–486.

Rand, M. and S. Catalano (2007, December). *Criminal Victimization, 2006.* NCJ 219413. Washington, DC: Bureau of Justice Statistics.

Rand, M. R. (1994, April). *Guns and Crime: Handgun Victimization, Firearm Self-Defense, and Firearm Theft.* NCJ 147003. Washington, DC: Bureau of Justice Statistics.

Rand, M. R. and C. M. Rennison (2005). Bigger is not necessarily better: An analysis of violence against women estimates from the National Crime Victimization Survey and the National Violence Against Women Survey. *Journal of Quantitative Criminology 21*(3), 267–291.

Rao, J. (2003). *Small Area Estimation.* New York: Wiley.

Rennison, C. M. (2003). *Violent Crime and Victims Gender, 2002: An Examination of Differences in Correlates of Male and Female Non-Fatal Violent Victimization in the State of Illinois, 2002.* Chicago: Illinois Criminal Justice Information Authority.

Rennison, C. M. and M. R. Rand (2007). Introduction to the National Crime Victimization Survey. See Lynch and Addington (2007), Chapter 2.

Robinson, L. (2007, March 21). Statement. Prepared testimony for the Subcommittee on Commerce, Justice, Science, and Related Agencies, Committee on Appropriations, U.S. House of Representatives.

Rosenfeld, R. (2007). Explaining the divergence between UCR and NCVS aggravated assault trends. See Lynch and Addington (2007), Chapter 9.

Rubin, M. (2007). *Maine Crime Victimization Report: Informing Public Policy for Safer Communities.* Edwin S. Muskie School of Public Service, University of Southern Maine. Portland, ME: Maine Statistical Analysis Center.

Savage, I. R. (1985). Hard-soft problems. *Journal of the American Statistical Association 80,* 1–7.

Singh, R. P. (1982, October 27). Investigation to determine the best reference period length for the National Crime Survey (NCS). Census Bureau memorandum, addressed to Gary M. Shapiro.

Sirken, M., T. Jabine, G. Willis, E. Martin, and C. Tucker (Eds.) (1999). *A New Agenda for Interdisciplinary Survey Research Methods: Proceedings of the CASM II Seminar*. Hyattsville, MD: National Center for Health Statistics.

Skogan, W. G. (1990). A review: The National Crime Survey redesign. *The Public Opinion Quarterly* 54(2), 256–272.

Smith, A. F. (2006). *Crime Statistics: An Independent Review*. Report of the Review Group. London: Home Office, United Kingdom.

Smith, M. D. (1987). Changes in the victimization of women: Is there a 'new female victim'? *Journal of Research in Crime and Delinquency* 24(4), 291–301.

Smith, M. D. and E. S. Kuchta (1993). Trends in violent crime against women, 1973–1989. *Social Science Quarterly* 74(1), 28–45.

Smith, S. K., G. W. Steadman, T. D. Minton, and M. Townsend (1999, May). *Criminal Victimization and Perceptions of Community Safety in 12 Cities, 1998*. Bureau of Justice Statistics and Office of Community Oriented Policing Services. NCJ 173940. Washington, DC: U.S. Department of Justice.

Spencer, B. D. (1982). Feasibility of benefit-cost analysis of data programs. *Evaluation Review* 6, 649–672.

Statistics Commission (2006, September). *Crime Statistics: User Perspectives*. Report No. 30. London: Statistics Commission, United Kingdom.

Steffensmeier, D. and M. D. Harer (1999). Making sense of recent U.S. crime trends, 1980 to 1996/1998: Age composition effects and other explanations. *Journal of Research in Crime and Delinquency* 36(3), 235–274.

Stohr, M. and S. Vazques (2001). *Idaho Crime Victimization Survey, 2000*. Meridian, ID: Idaho Statistical Analysis Center, Idaho State Police.

Sudman, S., N. M. Bradburn, and N. Schwarz (1996). *Thinking About Answers: The Application of Cognitive Processes to Survey Methodology*. San Francisco: Jossey-Bass.

Teske, R. H. and N. L. Lowell (1977). *Texas Crime Poll: Fall, 1977 Survey*. Huntsville, Texas: Survey Research Program, Criminal Justice Center, Sam Houston State University. Manuscript is undated and assumed to be 1977. Available: http://www.cjcenter.org/cjcenter/research/srp/cparchive/1977.pdf.

Thacher, D. (2004). The rich get richer and the poor get robbed: Inequality in U.S. criminal victimization, 1974–2000. *Journal of Quantitative Criminology* 20(2), 89–116.

Thaler, R. (1978). A note on the value of crime control: Evidence from the property market. *Journal of Urban Economics 5*, 137–145.

Tourangeau, R. and M. E. McNeeley (2003). Measuring crime and crime victimization: Methodological issues. See National Research Council (2003a), Chapter 2, pp. 10–42.

Tourangeau, R. and T. Smith (1996). Asking sensitive questions: The impact of data collection mode, question format, and question context. *Public Opinion Quarterly 60*, 275–304.

Turner, C., B. Forsyth, J. O'Reilly, P. Cooley, T. Smith, S. Rogers, and H. Miller (1998). Automated self-interviewing and the survey measurement of sensitive behaviors. In M. Couper, R. Baker, J. Bethlehem, C. Clark, J. Martin, W. Nicholls II, and J. O'Reilly (Eds.), *Computer-Assisted Survey Information Collection*. New York: Wiley.

U.S. Bureau of the Census (1968). *Report on National Needs for Criminal Justice Statistics*, pp. 53. In Bureau of Justice Statistics (1989).

U.S. Census Bureau (2003). *National Crime Victimization Survey: Interviewing Manual for Field Representatives*. NCVS-550. Washington, DC: U.S. Census Bureau.

U.S. Department of Health and Human Services (2000, November). *Healthy People 2010* (2nd ed.). With Understanding and Improving Health and Objectives for Improving Health. 2 vols. Washington, DC: U.S. Government Printing Office.

U.S. Office of Management and Budget (2000). *Statistical Programs of the United States Government: Fiscal Year 2001*. Washington, DC: Government Printing Office.

U.S. Office of Management and Budget (2006a, January). *Questions and Answers When Designing Surveys for Information Collections*. Washington, DC: Office of Information and Regulatory Affairs, U.S. Office of Management and Budget.

U.S. Office of Management and Budget (2006b, September). *Standards and Guidelines for Statistical Surveys*. Washington, DC: Statistical and Science Policy, Office of Information and Regulatory Affairs, U.S. Office of Management and Budget.

U.S. Office of Management and Budget (2006c). *Statistical Programs of the United States Government: Fiscal Year 2007*. Washington, DC: Government Printing Office.

U.S. Office of Management and Budget (2007). *Statistical Programs of the United States Government: Fiscal Year 2008*. Washington, DC: Government Printing Office.

Walker, A. (2006, January 23). Victim survey methodology: Mode, sample design and other aspects—results from the inventory of victimisation surveys. Working Paper Number 6, submitted by the Home Office, United Kingdom, for the Joint

United Nations Economic Commission for Europe and United Nations Office on Drugs and Crime Meeting on Crime Statistics, Vienna.

Walker, S. (1998). *Popular Justice: A History of American Criminal Justice* (2nd ed.). New York: Oxford University Press.

Warchol, G. (1998, July). *Workplace Violence, 1992–96.* NCJ 168634. Washington, DC: Bureau of Justice Statistics.

Wiersema, B. (1999). *Area-Identified National Crime Victimization Survey Data: A Resource Available Through the National Consortium on Violence Research.* NCOVR Census Center Working Paper 1. Pittsburgh, PA: National Consortium on Violence Research.

Woltman, H. and J. Bushery (1984). Summary of a study examining "differentially weighted estimates of annual victimization rates using a 12-month bounded reference period". See Lehnen and Skogan (1984), pp. 117–118.

Woltman, H., J. Bushery, and L. Carstensen (1984). Recall bias and telescoping in the National Crime Survey. See Lehnen and Skogan (1984), pp. 90–93.

Work, C. R. (1975). Objectives of the NCS and of victimization surveys. See National Research Council (1976b), pp. 219–228. Incorporated into memorandum to Charles Kindermann.

Wyoming Statistical Analysis Center (2004, Winter). First-ever Wyoming crime victimization survey. *Field Notes from the SRC.* Newsletter, Survey Research Center, University of Wyoming.

Ybarra, L. M. and S. L. Lohr (2002). Estimates of repeat victimization using the National Crime Victimization Survey. *Journal of Quantitative Criminology 18*(1), 1–21.

Young, M., T. DuPont-Morales, and J. Burns (1997). *Volume I: Pennsylvania Crime Victimization Survey, Executive Report.* The Center for Survey Research, Institute of State and Regional Affairs. Middletown, PA: Pennsylvania State University–Harrisburg.

Zawitz, M. W. (1995, July). *Guns Used in Crime: Firearms, Crime, and Criminal Justice.* NCJ 148201. Washington, DC: Bureau of Justice Statistics.

Zimring, F. E. and G. Hawkins (1995). *Incapacitation: Penal Confinement and the Restraint of Crime.* New York: Oxford University Press.

Appendixes

– A –

Findings and Recommendations

This appendix lists the panel's findings and recommendations for ease of reference.

Finding 3.1: As currently configured and funded, the NCVS is not achieving and cannot achieve BJS's legislatively mandated goal to "collect and analyze data that will serve as a continuous and comparable national social indication of the prevalence, incidence, rates, extent, distribution, and attributes of crime . . ." (42 U.S.C. 3732(c)(3)).

Recommendation 3.1: BJS must ensure that the nation has quality annual estimates of levels and changes in criminal victimization.

Recommendation 3.2: Congress and the administration should ensure that BJS has a budget that is adequate to field a survey that satisfies the goal in Recommendation 3.1.

Recommendation 3.3: BJS should continue to use the NCVS to assess crimes that are difficult to measure and poorly reported to police. Special studies should be conducted periodically in the context of the NCVS program to provide more accurate measurement of such events.

Recommendation 4.1: BJS should carefully study changes in the NCVS survey design before implementing them.

Recommendation 4.2: Changing from a 6-month reference period to a 12-month reference period has the potential for improving the precision per-unit cost in the NCVS framework, but the extent of loss of measurement quality is not clear from existing research based on the post-1992-redesign NCVS instrument. BJS should sponsor additional research—involving both experimentation as well as analysis of the timing of events in extant data—to inform this trade-off.

Recommendation 4.3: BJS should make supplements a regular feature of the NCVS. Procedures should be developed for soliciting ideas for supplements from outside BJS and for evaluating these supplements for inclusion in the survey.

Recommendation 4.4: BJS should maintain the core set of screening questions in the NCVS but should consider streamlining the incident form (either by eliminating items or by changing their periodicity).

Recommendation 4.5: BJS should investigate the use of modeling NCVS data to construct and disseminate subnational estimates of major crime and victimization rates.

Recommendation 4.6: BJS should develop, promote, and coordinate subnational victimization surveys through formula grants funded from state-local assistance resources.

Recommendation 4.7: BJS should investigate changing the sample design to increase efficiency, thus allowing more precision for a given cost. Changes to investigate include:

(i) changing the number or nature of the first-stage sampling units;

(ii) changing the stratification of the primary sampling units;

(iii) changing the stratification of housing units;

(iv) selecting housing units with unequal probabilities, so that probabilities are higher where victimization rates are higher; and

(v) alternative person-level sampling schemes (sampling or subsampling persons within housing units).

Recommendation 4.8: BJS should investigate the introduction of mixed mode data collection designs (including self-administered modes) into the NCVS.

Recommendation 4.9: The falling response rates of NCVS are likely to continue, with attendant increasing field costs to avoid their decline. BJS should sponsor nonresponse bias studies, following current OMB guidelines, to guide trade-off decisions among costs, response rates, and nonresponse error.

Recommendation 5.1: BJS should establish a scientific advisory board for the agency's programs; a particular focus should be on maintaining and enhancing the utility of the NCVS.

Recommendation 5.2: BJS should perform additional and advanced analysis of NCVS data. To do so, BJS should expand its capacity in the number and training of personnel and the ability to let contracts.

Recommendation 5.3: BJS should undertake research to continuously evaluate and improve the quality of NCVS estimates.

Recommendation 5.4: BJS should continue to improve the availability of NCVS data and estimates in ways that facilitate user access.

Recommendation 5.5: The Census Bureau and BJS should ensure that geographically identified NCVS data are available to qualified researchers through the Census Bureau's research data centers, in a manner that ensures proper privacy protection.

Recommendation 5.6: The Statistical Policy Office of the U.S. Office of Management and Budget is uniquely positioned to identify instances in which statistical agencies have been unable to perform basic sample or survey maintenance functions. For example, BJS was unable to update the NCVS household sample to reflect population and household shifts identified in the 2000 census until 2007. The Statistical Policy Office should note such breakdowns in basic survey maintenance functions in its annual report *Statistical Programs of the United States Government.*

Recommendation 5.7: Because BJS is currently receiving inadequate information about the costs of the NCVS, the Census Bureau should establish a data-based, data-driven survey cost and information system.

Recommendation 5.8: BJS should consider a survey design competition in order to get a more accurate reading of the feasibility of alternative NCVS redesigns. The design competition should be administered with the assistance of external experts, and the competition should include private organizations under contract and the Census Bureau under an interagency agreement.

Principal Findings and Recommendations of the National Research Council (1976b) Study

B–1 FINDINGS

1. The design of the NCS generally is consistent with the objective of producing data on trends and patterns of victimization for certain categories of crime.

2. Conceptual, procedural, and managerial problems limit the potential of the NCS, but the panel considers that, given sufficient support, the problems ought to be amenable to substantial resolution in the long run.

3. A major shift of resources to analytic and methodological research is essential in order for the NCS to yield data useful for policy formulation. This shift should be accompanied by the development of administrative mechanisms to enhance this large and complex series' capacity for self-correction.

4. The primary uses envisioned originally for the NCS were of a social and policy indicator nature. The panel agrees that a subsequent objective of producing operating intelligence for jurisdictions is inconsistent with the original purposes of the NCS and with the design informed by

those purposes, except insofar as operating intelligence is a by-product of understanding broad trends and patterns of victimization.

B–2 RECOMMENDATIONS

1. A substantially greater proportion of [Office of Justice Programs] resources should be allocated to delineation of product objectives, to managerial coordination, to data analysis and dissemination, and to a continuing program of methodological research and evaluation. We are concerned about the current balance between resources allocated to data collection and resources allocated to all other aspects of the victimization survey effort.

2. The staff providing managerial and analytic support for the NCS should be expanded to include the full-time efforts of at least 30 to 40 professional employees. Without this expansion, the NCS cannot be developed to achieve its potential for practical utility.

3. A coordinator at the Bureau of the Census should be appointed whose responsibilities would crosscut the various Census operations that support the NCS.

4. The staff that performs NCS analysis and report-writing functions, whether LEAA employees or otherwise, should have an active role in the management of the NCS. Specifically, the analytic staff should participate in the development of objectives for substantive reports and publication schedules. Once analytic plans are formulated, the analysis staff should have autonomy in specifying tabulations to be used in support of the analysis, and it should have direct access to complete NCS data files and to data processing resources. It should be the analytic staff's responsibility to formulate statistical or other criteria used in hypothesis testing. Finally, a feedback mechanism should be instituted through which the staff can influence decisions on the content of survey instruments, on field and code procedures, and on analytic and methodological research to be undertaken.

5. Resources now used for the nationwide household survey and for the independent city-level household surveys should be consolidated and used for carrying out an integrated national program. The integrated effort could produce not only nationwide and regional data, but, on the same timetable, estimates for separately identifiable Standard Metropolitan Statistical Areas (SMSAs) and for at least the five largest central cities within them. For some purposes, it would be practicable and perhaps useful to combine data for 2 or more years and to

show separate tabulations for a large number of cities and metropolitan areas.

6. A review and restatement of the objectives of the commercial surveys should be conducted and data collection should be suspended, except in support of experimental and exploratory review of these objectives.

7. Five percent of the NCS sample in the future should be available to interview in order to explore different forms and ordering of questions, and for pretesting possible new questions. . . .

8. Routine NCS tabulation should include results on the risk of victimization, where the unit of analysis is the surveyed individual, and that analysis of risk should be a significant part of NCS publications on a recurring basis. If the NCS data are coded and tabulated so as to yield a cumulative count of personal and household victim experiences of *all* surveyed respondents, analyses of multiple victimization, including events now excluded as "series" incidents, could and should be routine components of official publications.

9. A major methodological effort on optimum field and survey design for the NCS should be undertaken. Toward this goal, high priority should be given to research on the best combination of reference period, frequency of interview at an address, length of retention in the sample, and bounding rules. Part of the recommended research in this area should be a new reverse record check study in order to assess: (a) differential *degrees* of reporting for different types of victimizations and different classes of respondents, (b) problems of telescoping and decay, and (c) biases in the *misreporting* of facts.

10. Local interest in victimization patterns should be addressed through LEAA-Census joint development of a manual of procedures for conducting local area victimization surveys. The federal government should produce reports on the NCS that contain detailed analyses of patterns and trends of victimization so as to allow law enforcement personnel, the public, and policymakers to draw inferences that might be applicable to the issues with which they are concerned. Informing the public and their policymakers of the distribution and modifiability of risk should be the primary objective of the NCS.

SOURCE: National Research Council (1976b:3–5).

– C –

Procedures and Operations of the National Crime Victimization Survey

This summary derives heavily from the survey methodology documentation accompanying the Bureau of Justice Statistics (BJS) annual report *Criminal Victimization in the United States*, to which we refer readers for additional detail (see Bureau of Justice Statistics, 2006a).

C–1 SAMPLE DESIGN AND SIZE

C–1.a Sample Construction

The National Crime Victimization Survey (NCVS) "has a national sample of approximately 56,000 designated addresses located in approximately 673 primary sampling units throughout the United States" (Demographic Surveys Division, U.S. Census Bureau, 2007b). Specifically, the survey follows a stratified, multistage cluster sample design (Bureau of Justice Statistics, 2006a:130):

- *First stage:* The primary sampling units (PSUs) for the NCVS are "counties, groups of counties, or large metropolitan areas" that are grouped into strata.[1] Large PSUs (in population) are "self-

---

[1]The NCVS interviewer manual (U.S. Census Bureau, 2003:A1-9) implies that subcounty units—either the minor civil divisions that are functioning governmental units in some states

representing" (SR) because each is assigned to its own stratum and each is automatically selected for the sample. Decennial census data on "geographic and demographic characteristics" are used to group smaller PSUs into strata.[2] These smaller PSUs are "non-self-representing" (NSR) because only one PSU per stratum (selected with probability proportional to population size) is drawn for inclusion in the sample.

• *Second stage:* Within each sample PSU, "clusters of approximately four housing units or housing unit equivalents" are drawn from each of four nonoverlapping frames: (1) the unit frame, a listing of housing units for the PSU from the decennial census Master Address File (MAF); (2) the group quarters frame, a listing of nonhousehold units such as college dormitories and rooming houses derived from the decennial census roster of such places; (3) the area frame, consisting of census blocks; and (4) the permit frame, a listing of addresses compiled from building permit data. The use of the permit data allows for housing constructed after the most recent decennial census to be included in the NCVS.

For NCVS data collected in 2005, the sample design consisted of 93 SR PSUs and 110 NSR strata. The original design based on 1990 census files included 152 NSR strata, but 42 were eliminated due to a sample cut in 1996 (see Box 1-2).

The various units, strata, and frames used in the sample design of the NCVS are keyed to the decennial census; however, the transition to files based on a new decennial census is not as rapid as might be imagined. Sample based on addresses from the 2000 census Master Address File only began to be phased in starting in 2005; as of 2007, the NCVS remains a hybrid of 2000- and 1990-census-based sample. Likewise, both 1980- and 1990-based sample were used through 1997, with 1990-only sample only beginning in 1998.

With respect to the group quarters coverage of the NCVS sample, it is important to note that BJS and the Census Bureau exclude some major types of nonhousehold, group quarters types from eligibility in the survey. In particular, "institutionalized persons, such as correctional facility inmates," are not included in the NCVS, nor are seaborne personnel on merchant vessels or armed forces personnel living in barracks.

(e.g., townships) or the similarly sized census county divisions that the Census Bureau defines for other states—may also be grouped into strata. Hence, PSUs need not be made up of whole counties. This is particularly so in New England and Hawaii, where stratification PSUs "could consist of one or more" of these subcounty areas.

[2] "These characteristics include geographic region, population density, rate of growth, population, principal industry, and type of agriculture" (U.S. Census Bureau, 2003:A1-9).

C–1.b   Person-Level Eligibility for Inclusion in the NCVS

In addition to the restriction of some group quarters types from NCVS eligibility—thus excluding people like jail or prison inmates—the NCVS also restricts the coverage of the survey by age. Within units selected for the sample, NCVS interviewers are directed to obtain reports for only those individuals age 12 or older.

Furthermore, "U.S. citizens residing abroad and foreign visitors to this country" are not supposed to be included in the NCVS.

The youngest of NCVS-eligible persons—12- and 13-year-olds—are one of three cases in which proxy reporting (and not a direct interview) is permitted; if the household respondent (see Section C–3.a) insists that the interviewer not directly question a 12- or 13-year-old, questionnaire information may be taken from that household member. The other two cases in which proxy reporting is allowed are "temporarily absent household members and persons who are physically or mentally incapable of granting interviews"; in the latter instance, a person who is not a member or usual resident of the household (e.g., a professional caregiver) may provide the requested information.[3]

C–1.c   Sample Size Over Time

Table C-1 shows the number of sample households and the number of persons contacted in those households for the most recent years for which NCVS data are available. The table also illustrates the decline in NCVS sample size over time. When the National Crime Survey began in 1972, it reached a household sample of 72,000 households; by 2005, that sample had nearly been halved to 38,600. Some of the sample size cuts contributing to this decline are listed in Box 1-2.

The combination of a declining sample size and less crime to measure—overall, "the rate of crime remains at the lowest levels in the past thirty years" (Catalano, 2006:3)—makes it extremely difficult to discern year-to-year annual change, even in aggregate measures like the rate of all violent crime. The violent crime rates for 1993–2005 are illustrated in Figure C-1, along with boxes showing the 95 percent confidence intervals associated with those rates; significant annual change differences have been rare in the past decade. For this reason, the BJS *Criminal Victimization* bulletins (e.g., Catalano, 2006) make comparisons based on 2-year groups of data.

---

[3]Interpreters and signers are permitted to respond for persons who do not understand English or who are deaf, respectively, but these are not considered proxy respondents.

Table C-1   Number of Households and Persons Interviewed
by Year, 1996–2005

Year	Households Sample Size	Response Rate	Persons Sample Size	Response Rate
1996	45,000	93	85,330	91
1997	42,910	95	79,470	90
1998	43,000	94	78,900	89
1999	43,000	93	77,750	89
2000	43,000	93	79,710	90
2001	44,000	93	79,950	89
2002	42,000	92	76,050	87
2003	42,000	92	74,520	86
2004	42,000	91	74,500	86
2005	38,600	91	67,000	84

NOTE: These sample sizes correspond to the number of separate
households and persons designated for contact in a particular year.
Participation rates for a particular year would be roughly double these,
accounting for two interviews with sample addresses in the same year.

SOURCE: Bureau of Justice Statistics (2006a).

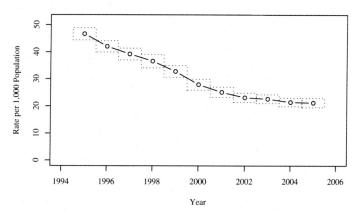

Figure C-1   Year-to-year change in NCVS violent crime victimization rate,
1993–2005

NOTE: Boxes denote the 95 percent confidence interval for each year's victimization rate.
Rates are for violent victimizations per 1,000 population for persons age 12 and over.

SOURCE: Bureau of Justice Statistics; see Catalano (2006).

C–2 ROTATING PANEL DESIGN

The NCVS follows a rotating panel design: its sample of addresses, drawn from the four different sampling frames described above, is divided into six equally sized "rotation groups," and each rotation group is further divided into six "panels." One panel from each of the rotation groups is designated for interviewing each month.

NCVS interviews are scheduled at 6-month intervals, and each household remains in the sample for 3.5 years or a total of 7 interviews. A new rotation group is added into the sample every 6 months, to replace the rotation group that is exiting the sample because its 3.5-year eligibility is complete. In each interview, Census Bureau field representatives are directed to identify a "household respondent," who is the first person to be interviewed and the only person to be asked questions about crimes against the household as a whole. In the second through seventh interviews conducted with the same household, the interviewers are encouraged to use the same person as the household respondent, if possible. The household respondent is discussed further in Section C–3.a.

C–2.a   Reference Period

The NCVS uses a 6-month reference period; that is, it asks its respondents to list and provide information on victimization incidents they experienced within the 6-month window before the NCVS interview. The reference period used for an interview in a particular month is graphically illustrated in Table C-2.

The NCVS Resource Guide maintained by the National Archive of Criminal Justice Data[4] summarizes the basic issues and inherent trade-offs in designating a particular reference period as follows:

> Generally, respondents are able to recall more accurately an event which occurred within three months of the interview rather than one which occurred within six months; they can recall events over a six-month period more accurately than over a 12-month period. However, a shorter reference period would require more field interviews per year, increasing the data collection costs significantly. These increased costs would have to be balanced by cost reductions elsewhere (sample size is often considered). Reducing sample size however, reduces the precision of estimates of relatively rare crimes. In light of these trade-offs of cost and precision, a reference period of six months is used for the NCVS.

---

[4]See http://www.icpsr.umich.edu/NACJD/NCVS/ [6/4/07].

Table C-2 Month of Incident by Month of Interview in the Current NCVS Sample Design

Month of Incident	Jan	Feb	Mar	Apr	May	Jun	Jul	Aug	Sep	Oct	Nov	Dec	Jan	Feb	Mar	Apr	May	Jun
	t	t	t	t	t	t	t	t	t	t	t	t	t+1	t+1	t+1	t+1	t+1	t+1
Jul t−1	X																	
Aug t−1	X	X																
Sep t−1	X	X	X															
Oct t−1	X	X	X	X														
Nov t−1	X	X	X	X	X													
Dec t−1	X	X	X	X	X	X												
Jan t		X	X	X	X	X	X											
Feb t			X	X	X	X	X	X										
Mar t				X	X	X	X	X	X									
Apr t					X	X	X	X	X	X								
May t						X	X	X	X	X	X							
Jun t							X	X	X	X	X	X						
Jul t								X	X	X	X	X	X					
Aug t									X	X	X	X	X	X				
Sep t										X	X	X	X	X	X			
Oct t											X	X	X	X	X	X		
Nov t												X	X	X	X	X	X	
Dec t													X	X	X	X	X	X

NOTES: Xs denote months in the 6-month reference period, for which any victimization incidents are supposed to be reported at the time of interview. Grey shading indicates the months that are combined to produce *collection-year* estimates for year t (collecting all interviews conducted in year t, which can include reports of incidents occurring in the last half of year $t − 1$). The ruled box denotes months that would be used for *data-year* estimates for year t (collecting interviews where incidents occurred in year t).

SOURCE: Bureau of Justice Statistics (2000:63).

C–2.b   Mode of Survey Contact Over Repeated Interviews

When the NCVS began, it relied strictly on face-to-face visits from Census Bureau field staff to sample households conducting pencil-and-paper interviews. Beginning in 1980, BJS and the Census Bureau began to shift toward increased use of telephone interviewing in the interest of reducing survey costs. It was deemed essential—and remains a requirement today— that the first interview with a sample household be conducted through a personal interview. But, starting in 1980, the use of the telephone in every other interview (2, 4, and 6) was encouraged. As described in Box 1-2, this was later revised to encourage even wider use of telephone interviewing, with only interviews 1 and 5 slated for personal visits. By 2003, "subsequent NCVS interviews" after the initial interview were to be carried out by phone "whenever possible" (U.S. Census Bureau, 2003:A1-11); interviewers were advised that "most of your NCVS interviews will be by telephone" because "telephone interviews are more cost effective" (U.S. Census Bureau, 2003:A5-3). Further, the restriction that the first-time-in-sample interview be conducted in person applies only to the household respondent (see Section C–3.a); "any other eligible household members who are not available during [the interviewer's initial visit] can be interviewed by telephone" (U.S. Census Bureau, 2003:A5-3). Accordingly, some NCVS respondents may never be interviewed face-to-face in a personal visit during the 3.5 years a household remains in sample.

About 30 percent of interviews with sample households were designated for completion by computer-assisted telephone interviewing (CATI) from Census Bureau CATI centers in Hagerstown, Maryland, and Tucson, Arizona (U.S. Census Bureau, 2003:A1-11). By 2007, BJS concluded that the expected cost savings from CATI interviewing from the Census Bureau call centers had not been realized. As part of a package of revisions to cover NCVS costs for three years, BJS and the Census Bureau decided to suspend the use of the Census Bureau CATI centers for the NCVS. However, this covers only the formal CATI interviews from the designated call centers, which is to say that "informal" telephone interviewing—with Census Bureau field representatives calling households in their designated workloads to complete interviews by phone—is still employed (and encouraged) for interviews 2–7 with sample households.

Prior to the first-time-in-sample interview, conducted in person, an introductory letter is sent by the Census Bureau to each sample housing unit, advising household members of their selection for the sample and providing the household with the address and phone number of the appropriate Census Bureau regional office. A similar introductory letter "is usually mailed before each subsequent enumeration period" (U.S. Census Bureau, 2003:A2-

3). Unlike the decennial census (and the new American Community Survey conducted by the Census Bureau), responses to the NCVS are not required by law, and households are advised of the voluntary nature of the survey, in compliance with federal privacy laws.

C–2.c Bounding

A common concern of researchers employing reference periods in retrospective surveys is telescoping. Telescoping refers to a respondent's misspecification of when an incident occurred in relation to the reference period. For example, telescoping occurs if a respondent is asked about victimizations within the last six months and erroneously includes a victimization that occurred eight months ago. Telescoped events that actually occurred prior to the reference period can be minimized at the time of the first interview by a technique known as bounding. Bounding is achieved by comparing incidents reported in an interview with incidents reported in a previous interview and deleting duplicate incidents that were reported in the current reference period. In the National Crime Survey (NCS) and NCVS designs, each visit to a household is used to bound the next one by comparing reports in the current interview with those given six months prior. When a report appears to be a duplicate, the respondent is reminded of the earlier report and asked if the new report represents the incident previously mentioned or if it is different. The first interview at a household entering the sample is unbounded, and data collected at these interviews were not included in NCS and NCVS estimates until very recently.

Skogan (1990:262) notes that the use of the bounding interview emerged as a high priority from the initial pilot studies that preceded the full National Crime Survey:

> One of the clearest methodological findings of the pilot studies . . . was that people draw incidents into the temporal window that they are *supposed* to describe ("the past six months") when in fact those incidents occurred outside of the time frame. The effect of these out-of-range incidents is very large, increasing the victimization count by between 40% and 50% depending on the type of crime; the inflation is greatest for violent crimes and those (often more serious) that were reported to police, and it is smallest for simple thefts.

Hence, withholding the first interview for use in bounding acted as a safeguard for including these out-of-range incidents.

However, the NCVS sample is a rotating panel of addresses rather than people. As such, "movers"—addresses whose occupants change during the time that the address is in the sample—pose a challenge for the practice of withholding a bounding interview. Historically, the NCVS treated the problem in the simplest and most cost-effective way: include interviews 2–7 in the estimates, regardless of whether a bounding interview exists for the specific household at an address. That is, if a sample address changes hands between interview periods, the first interview with the new household members is unbounded but included in NCVS estimates. This use of unbounded interviews for mover households "inflat[es] estimates of the victimization rate. The number of respondents involved is substantial; the yearly attrition rate from the NCS sometimes approaches 20%, although that figure fluctuates" Skogan (1990:262). Cantor and Lynch (2000:119) cite the use of unbounded data from movers as an example of an inconsistency in treatment within the NCVS that may be correlated with measurement error. In addition to inflated reports of victimization, they note (citing Biderman and Cantor, 1984) that "those [people who are] most likely to move have higher victimization rates," so that the survey "will overestimate the relationship between mobility (and its correlates) with victimization."

Alternative techniques for handling movers have been considered from time to time. In particular, the consortium that redesigned the survey over the 1980s (resulting in the new NCVS in 1992) considered "a scheme for retaining individuals who move in the sample, tracking them over time for up to several more years; the design was akin to that now utilized by the Census Bureau's Survey of Income and Program Participation (SIPP)" (Skogan, 1990:263).

NCVS practices for bounding interviews and mover households remained the same until enactment of a set of cost-cutting measures in 2007 (for application to 2006 data). Under the new plan, unbounded first-time-in-sample interviews were to be included in NCVS estimates; this change was accompanied by a corresponding reduction in overall sample size.

C–3 STRUCTURE OF THE NCVS INSTRUMENT AND INTERVIEW

In the sections that follow, we generally describe the features of the NCVS instrument and the progress of an NCVS interview through reference to the most recent versions of the survey in paper-and-pencil format. After several years of work, the NCVS is now fully implemented through computer-assisted means, with interviewers using laptop computers to administer the interviews (or, until their abandonment, with interviews conducted from Census Bureau–operated telephone call centers). Hence, some of the terminology (e.g., Control Card) and naming conventions (e.g., a NCVS-2 form) is not in current usage, but they remain useful in describing the general structure of the survey.

C–3.a   Household and Individual Respondents

Within each household contacted at a sample address, one person is designated the *household respondent*; the other eligible persons for interviewing are termed *individual respondents*. Interviewers are urged "to find the **most knowledgeable** household member who is at least 18 years of age" for selection as the household member. Typically, this is "one of the persons who owns or rents the home," and as such is also the *reference person* for the household (for determining each individual respondent's relationship to the household). In the subsequent interviews at the same address, interviewers are asked to try to use the same person as the household respondent (U.S. Census Bureau, 2003:A2-10).

The household respondent is the first person interviewed and is the only person in the household who is asked about "thefts of certain kinds of things that are considered the common property of the household," including questions about "burglary, motor vehicle theft, and the theft of specific household property such as plants or lawn furniture" (Cantor and Lynch, 2000:95). Although this approach avoids duplication and reduces the overall reporting burden for the household, Cantor and Lynch (2000:119) (citing Biderman et al., 1985) suggest that this household screener approach also "reveals more crimes against individuals as well. . . . The selection of household [respondents] is negatively correlated with victim risk (the household member who tends to stay home is most likely to be selected as the household [respondent]). This depresses relationships associated with risk."

C–3.b   Control Card

As described in NCVS technical documentation (Bureau of Justice Statistics, 2007a):

> The Control Card is the basic administrative record for each sample unit. It contains the address of each sample unit and the basic household data, such as the names of all persons living there and their age, race, sex, marital status, education and the like. Household income, tenure of the unit and pertinent information about non-interviews are also included on the Control Card. The Control Card serves as a record of visits, telephone calls, interviews, and non-interview reasons. The Control Card information is updated, as needed, during each visit to the housing unit, except for questions about educational attainment, income, and tenure, which are only asked every other visit.

The Control Card also serves as a vehicle for bounding interviews (see Section C–2.c), in addition to the application of weights in generating estimates. At the end of an Incident Report (described below in Section C–3.d), basic summary information about the incident is summarized and compared against entries on the Control Card. The NCVS interviewing manual (U.S. Census Bureau, 2003:C1-36) advises interviewers that:

> If you suspect that a duplication has occurred, tactfully ask the respondent whether the incidents are the same (*for example, same time, place, and circumstances*) or separate crime incidents (*for example, same place and circumstances, but different times*).

If the newly reported incident is determined to be the same as a Control Card entry, then it is marked as "out of scope" for further processing.

C–3.c Screening Questions (NCVS-1)

The core of the Basic Screen Questionnaire—referred to, from its paper incarnation, as form NCVS-1—is a set of screening questions. Interviewers are instructed to tell respondents that "I'm going to read some examples that will give you an idea of the kinds of crimes this study covers. As I go through them, tell me if any of these happened to you in the last six months, that is, since *[Specific Reference Date]*." Each subsequent screening question (or "screener," for short) consists of a fairly detailed description of a particular victimization type. The respondent is asked, "Did any incidents of this type happen to you?" If the answer is "yes," the respondent is asked, "How many times?" and is allowed to briefly describe the incident. After the bank of screening questions is complete, interviewers are directed to go through an Incident Report—described in the next section—for each category for which the number of reported incidents is greater than zero. This process is repeated for each person age 12 or older in the household.

As of the 2005 version of the NCVS instrument, the basic screener questions (as administered to the household respondent, and so including the questions about crimes against the household) were as follows:

- Q36a: Was something belonging to YOU stolen, such as –

  (a) Things that you carry, like luggage, a wallet, purse, briefcase, book –
  (b) Clothing, jewelry, or cellphone –
  (c) Bicycle or sports equipment –
  (d) Things in your home – like a TV, stereo, or tools
  (e) Things outside your home such as a garden hose or lawn furniture –
  (f) Things belonging to children in the household –
  (g) Things from a vehicle, such as a package, groceries, camera, or CDs – OR
  (h) Did anyone ATTEMPT to steal anything belonging to you?

- Q37a: [Other than any incidents already mentioned,] has anyone –

(a) Broken in or ATTEMPTED to break into your home by forcing a door or window, pushing past someone, jimmying a lock, cutting a screen, or entering through an open door or window?

(b) Has anyone illegally gotten in or tried to get into a garage, shed, or storage room? OR

(c) Illegally gotten in or tried to get into a hotel or motel room or vacation home where you were staying?

- *Following a question on the total number of motor vehicles owned by the household,* Q39a: During the last 6 months, (other than any incidents already mentioned,) (was the vehicle/were any of the vehicles) –

(a) Stolen or used without permission?

(b) Did anyone steal any parts such as a tire, car stereo, hubcap, or battery?

(c) Did anyone steal any gas from (it/them)? OR

(d) Did anyone ATTEMPT to steal any vehicle or parts attached to (it/them)?

- Q40a: (Other than any incidents already mentioned,) since *[Specific Reference Date]*, were you attacked or threatened OR did you have something stolen from you –

(a) At home including the porch or yard –

(b) At or near a friend's, relative's, or neighbor's home –

(c) At work or school –

(d) In places such as a storage shed or laundry room, a shopping mall, restaurant, bank, or airport –

(e) While riding in any vehicle –

(f) On the street or in a parking lot –

(g) At such places as a party, theater, gym, picnic area, bowling lanes, or while fishing or hunting – OR

(h) Did anyone ATTEMPT to attack or ATTEMPT to steal anything belonging to you from any of these places?

- Q41a: (Other than any incidents already mentioned,) has anyone attacked or threatened you in any of these ways (Exclude telephone threats) –

(a) With any weapon, for instance, a gun or knife –

(b) By something thrown, such as a rock or bottle –

(c) With anything like a baseball bat, frying pan, scissors, or stick –

(d) Include any grabbing, punching, or choking,

(e) Any rape, attempted rape or other type of sexual attack –

(f) Any face to face threats – OR

(g) Any attack or threat or use of force by anyone at all? Please mention it even if you are not certain it was a crime.

- Q42a: People often don't think of incidents committed by someone they know. (Other than any incidents already mentioned,) did you have something stolen from you OR were you attacked or threatened by (Exclude telephone threats) –

    (a) Someone at work or school –
    (b) A neighbor or friend –
    (c) A relative or family member –
    (d) Any other person you've met or known?

- Q43a: Incidents involving forced or unwanted sexual acts are often difficult to talk about. (Other than any incidents already mentioned,) have you been forced or coerced to engage in unwanted sexual activity by –

    (a) Someone you didn't know before –
    (b) A casual acquaintance – OR
    (c) Someone you know well?

- Q44a: During the last 6 months, (other than any incidents already mentioned,) did you call the police to report something that happened to YOU which you thought was a crime? *If the answer is "yes," a brief description is sought. The interviewer is then instructed to review this description to see whether it adequately answers the question:* Were you (was the respondent) attacked or threatened, or was something stolen or an attempt made to steal something that belonged to you (the respondent) or another household member? *If the interviewer is unsure, they are told to directly ask the respondent this follow-up question; otherwise, they may mark "yes" or "no" without asking.*

- Q45a: During the last 6 months, (other than any incidents already mentioned,) did anything which you thought was a crime happen to YOU, but you did NOT report to the police? *Like Q44a, a description is sought if the answer is yes. Based on that description, the interviewer either marks "yes" or "no" to or directly asks the question:* Were you (was the respondent) attacked or threatened, or was something stolen or an attempt made to steal something that belonged to you (the respondent) or another household member?

- Q46a: Now I'd like to ask about ALL acts of vandalism that may have been committed during the last 6 months against YOUR household. Vandalism is the deliberate, intentional damage to or destruction of household property. Examples are breaking windows, slashing tires, and painting graffiti on walls. Since *[Specific Reference Date]*, has anyone intentionally damaged or destroyed property owned by you or someone else in your household? (EXCLUDE any damage done in conjunction with incidents already mentioned.) *If "yes," the respondent is asked a series of follow-ups, including the type of property vandalized and whether the damage was more or less than $100, before the "How many times?" question.*

[The 2005 questionnaire also included a 7-part screening question regarding hate crimes; answers of "yes" to particular queries in that screener resulted in the completion of a separate incident report, not the standard NCVS Incident Report.]

Recognizing that the multiple-example structure of the screening questions "may prompt some respondents to give you an answer before you finish reading each subcategory," Census Bureau interviewers are told that the bureau "would prefer that you finish reading each subcategory" before an answer and, "even if you get interrupted, you must read each and every subcategory in its entirety" (U.S. Census Bureau, 2003:A2-33).

C–3.d Incident Report (NCVS-2)

After the screening questions are complete and before a respondent is asked a set of questions about their employment, the NCVS interviewing protocol is to complete an Incident Report for each victimization incident counted by the screening questions. "For example, if a respondent said that his pocket was picked once and he was beaten up twice, three Crime Incident Reports, one for each separate incident, are completed" (Bureau of Justice Statistics, 2007a:10).

An important exception to the general rule of one Incident Report per incident is what the NCVS considers *series victimizations*, which is a set of incidents meeting three conditions (Bureau of Justice Statistics, 2007a:11):

(1) Incidents must be of the same type or very similar in detail ("similar" is not defined in any more explicit detail in either the instrument or the interviewer manual).

(2) There must be at least 6 incidents in the series within the 6-month reference period; this threshold was changed in 1993, prior to which 3 incidents could define a series.

(3) The respondent must not be able to recall dates and other details of the individual incidents—that is, detail that would be sufficient to "complete most items" on the Incident Report (U.S. Census Bureau, 2003:C3-4) and that victimization types can be correctly coded—well enough to report them separately.

The survey documentation further notes that "interviewers are instructed to try through probing to get individual reports whenever possible and only accept series reports as a last resort" (Bureau of Justice Statistics, 2007a:11). In cases in which all three conditions apply, the respondent is asked to complete the incident form by providing the details of only the most recent incident in the series.

As of the paper questionnaire used in 2005, the NCVS-2 Incident Report included 173 numbered questions. However, the exact form of the interview and number of questions asked by the Incident Form varies based

Table C-3 Number of Interviews Completed by NCVS Sample Households, 1995–1999

Number of Completed Interviews	N	Percentage
0	3,552	11.44
1	1,870	6.02
2	1,629	5.25
3	1,420	4.57
4	1,451	4.67
5	2,484	8.00
6	4,657	15.00
7	13,985	45.04

NOTES: Tabulations from special longitudinally linked file prepared by the U.S. Census Bureau, following 31,048 sample households (sample J19, rotations 2, 3, and 4) from first-time-in-sample through 7 enumeration periods (3.5 years). The file was created based on quarterly edited files from quarter 3, 1995, to quarter 4, 1999. Of the households, 4,265 (13.74 percent) were interviewed on *both* the first and seventh occasion, with some mix of interviews and noninterviews in periods 2–6.

SOURCE: Demographic Surveys Division, U.S. Census Bureau (2007a:Table 2).

on the information reported, as "skip sequences" in the questionnaire route the interview past sections that would be irrelevant based on the provided information.

C-3.e   Attrition in the NCVS

Tabulations from a special longitudinally linked data file from the late 1990s (see Table C-3) suggest that just under half (45.04 percent) of households in the NCVS sample complete the full set of seven interviews.

C-4   SUPPLEMENTS TO THE NCVS

Topical supplements that have been added to the NCVS have included (Demographic Surveys Division, U.S. Census Bureau, 2007a):

- School Crime Supplement (1989, 1995, 1999, 2001, 2005, 2007), on school safety issues;

- Police-Public Contact Survey (1996 pilot, 1999, 2002, 2005; planned for 2008), on the nature of contacts with the police;

- National Survey of Crime Severity (1977), on public perceptions of crime severity;

- Victim Risk Supplement (1984), on crime prevention measures taken by household respondents;

- Workplace Risk Supplement (2002), on risk factors contributing to nonfatal violence in the workplace;

- Supplemental Victimization Survey (2006), on stalking or harassing behavior; and

- Identity Theft Supplement (planned for 2008), on the incidence and prevalence of identity theft.

The Census Bureau (Demographic Surveys Division, U.S. Census Bureau, 2007b:51) notes that the National Center for Education Statistics "bears all costs" of the School Crime Supplement.

> The supplement contains questions on preventative measures employed by the school to deter crime; students' participation in extracurricular activities; transportation to and from school; students' perception of rules and equality in school; bullying and hate crime in school; the presence of street gangs in school; availability of drugs and alcohol in the school; attitudinal questions relating to the fear of victimization in school; access to firearms; and student characteristics such as grades received in school and postgraduate plans.

Specifically, for the 2005 implementation of the supplement, "approximately 10,000 households containing approximately 11,600 respondents were eligible for the supplement from January through June 2005." The school crime questions were administered to all individual respondents in the sample households who were between ages 12 and 18, "who were enrolled in primary or secondary education programs leading to a high school diploma, and who were enrolled in school sometime during the six months prior to the interview." The School Crime Supplement is distinct from the Schools Survey on Crime and Safety, sponsored by the National Center for Education Statistics and conducted by the Census Bureau, which is a periodic, nationally representative cross-sectional survey of public elementary and secondary schools (administered to principals).

Another example of an NCVS supplement, the Supplemental Victimization Survey, was conducted from January through June 2006 (with no current plans for repetition). It was sponsored by the Office of Violence Against Women of the U.S. Department of Justice. Of the 56,000 NCVS households in that year, "approximately 42,700 households containing approximately 79,000 respondents were eligible for the supplement. The U.S. Census Bureau interviewers administered the supplemental interview to all people within these households who are 18 years of age or older and whose NCVS interview was conducted by self-response" (Demographic Surveys Division, U.S. Census Bureau, 2007b:54).

– D –

The Uniform Crime Reporting Program

Coordinated by the Federal Bureau of Investigation (FBI), the Uniform Crime Reporting (UCR) program is a cooperative program of law enforcement agencies that produces aggregate data on crimes reported to police. UCR data collection began in January 1930, drawing information from 400 U.S. cities; as of 2004, 17,000 law enforcement agencies participate in UCR (Federal Bureau of Investigation, 2004:Foreword). As it has evolved, the UCR program consists of two major systems: the long-established Summary Reporting System (SRS) and a newer, more detailed reporting system—the National Incident Based Reporting System (NIBRS)—that is poised to eventually supplant the SRS but has been slow to develop. We discuss these two component systems in turn.

The major features of the UCR compared with the National Crime Victimization Survey (NCVS) are summarized in Table D-1.

D–1 SUMMARY REPORTING SYSTEM

D–1.a Index Crimes

The core content of the Uniform Crime Reporting program inherits directly from the work of a Committee on Uniform Crime Records convened by the International Association of Chiefs of Police (IACP) in 1927. "Recognizing a need for national crime statistics," that committee "evaluated various crimes on the basis of their seriousness, frequency of occurrence, perva-

Table D-1   National Data Sources Related to Crime Victimization in the United States

Data Characteristics	NCVS	UCR Summary	NIBRS
Target population	Noninstitutionalized persons age 12 and older in the United States	Crime incidents occurring in the United States	Crime incidents occurring in the United States
Unit of observation	Individual	Law enforcement agency	Crime incident
Estimated coverage	Nationally representative sample	94.2 percent of United States population covered by agencies active in UCR reporting	Approximately 25 percent of United States population covered by agencies reporting in NIBRS format
Types of victimization covered			
Criminal Homicide	No	Yes	Yes
Other Index Crimes	Yes	Yes	Yes
Geographic areas identified			
Region	Yes	Yes	Yes
State	Yes	Yes	Yes
County	Yes	Yes	Yes
Census Tract	Yes	No	No
Demographic coverage			
Age	Yes	No	No
Race	Yes	No	Yes
Sex	Res	No	Yes
Ethnicity	Res	No	Yes
Vulnerable groups			
Children	12 & older	No	Yes
Immigrants (native born)	No	No	No
Disabled (learning disability only)	No	No	No
Elderly	Yes	No	No
Timeliness of data availability			
Time between reference period and data availability			
Pre-announced schedule	Yes	Yes	Yes
Fixed schedule	Yes	Yes	Yes
Accuracy and quality			
Sampling error	Routinely estimated	Unmeasured	Unmeasured
Other errors (nonsampling)	No ongoing evaluation	Unknown	Unknown

siveness in all geographic areas, and likelihood of being reported to law enforcement" (Federal Bureau of Investigation, 2004:2). Although the labels have changed slightly, the seven crimes identified by the 1927 IACP committee remain the focus of today's Uniform Crime Reports and are known as "Part I offenses." Three of these are crimes against persons—criminal homicide, forcible rape, and aggravated assault—and four are crimes against property: robbery, burglary, larceny-theft, and motor vehicle theft. The only substantive change to this list of Part I offenses was made in 1978, when legislation directed that arson be designated a Part I offense; however, arson continues to be reported on a separate form rather than the standard "Return A" used to report the other Part I offenses.

The Part I offenses are also known as "index crimes" because they are used to derive a general, national indicator of criminality—the national Crime Index. The index—first computed and reported in 1958—consists of the sum of the seven original Part I offenses, except that larceny is restricted to thefts of over $50.

D–1.b   Hierarchy Rule

The general order in which the Part I offenses are listed is not accidental. Instead, with some interleaving, the listing of offenses defines a strict hierarchy that agencies are asked to follow in coding offenses. In descending order, the UCR hierarchy by Part I offense and suboffense is as follows:

1. Criminal homicide

    a. Murder and nonnegligent manslaughter

    b. Manslaughter by negligence

2. Forcible rape

    a. Rape by force

    b. Attempts to commit forcible rape

3. Robbery

    a. Firearm

    b. Knife or cutting instrument

    c. Other dangerous weapon

    d. Strong-arm (hands, fists, feet, etc.)

4. Aggravated assault

    a. Firearm

    b. Knife or cutting instrument

    c. Other dangerous weapon

    d. Strong-arm (hands, fists, feet, etc.)

5. Burglary

    a. Forcible entry

    b. Unlawful entry (no force)

    c. Attempted forcible entry

6. Larceny-theft (except motor vehicle theft)

7. Motor vehicle theft

    a. Autos

    b. Trucks and buses

    c. Other vehicles

8. Arson

a.–g. Structural

h.–i. Mobile

    j. Other

For purposes of UCR collection, the FBI directs that multiple-offense situations—incidents in which more than one crime is committed simultaneously—are to be handled by "locat[ing] the offense that is highest on the hierarchy list and scor[ing] that offense involved and not the other offense(s)" (Federal Bureau of Investigation, 2004:10). However, three major exceptions to the general hierarchy rule are defined. First, motor vehicle theft—as a special class of larceny, generally—can outrank larceny; hence, the theft of a car with valuables inside it would be coded as a motor vehicle theft (trumping the classification as larceny) even if the vehicle is subsequently recovered but the valuables are not. Arson is also a special case because it is reported on a separate form from the other Part I offense: multiple-offense crimes involving arson can include *two* reported Part I offenses, the arson tally on the separate schedule and the highest-ranking Part I offense under the usual rule reported on Return A. The third exception to the hierarchy rule is justifiable homicide, "defined as and limited to the killing of a felon by a police officer in the line of duty [or] the killing of a felon, during the commission of a felony, by a private citizen." By this definition, justifiable homicide necessarily "occurs in conjunction with other offenses"; those offenses are the ones to be considered in classifying the incident.

Addington (2007:229) notes that the NCVS uses a "seriousness hierarchy"—comparable to the UCR hierarchy rule—for classification of events in incident count tabulations. However, the NCVS practice differs from the UCR rule in that "the NCVS collects and preserves information for each crime occurring in the incident, which enables researchers to create their own classification scheme;" in comparison, application of the UCR hierarchy rule collapses incidents involving several crime types to record just one type, losing the full incident detail.

D–1.c Supplemental Reports

In the Summary Reporting System, participating agencies are asked to report counts of all Part I offenses known to law enforcement on a standard, monthly form known as Return A. However, Return A is not the only data collection requested by the FBI. The Summary Reporting System also asks participating agencies to complete additional forms, at various intervals:

- *Age, race, and sex arrest data:* On a monthly basis, agencies are asked to provide counts of completed arrests by the age, race, and sex of the arrestee(s). Specifically, the age, sex, and race breakdowns are required for arrests for each of the Part II offenses, making these data the UCR's only systematic source of information on these offenses as well as the only source of offender attributes.[1]

- *Law enforcement officers killed and assaulted:* Data collected annually since 1972.

- *Hate crime statistics:* The Hate Crime Statistics Act of 1990 led to the collection of a variable on "bias motivation in incidents in which the offense resulted in whole or in part because of the offender's prejudice against a race, religion, sexual orientation, or ethnicity/national origin" (Federal Bureau of Investigation, 2004:3). The scope of hate crimes reported in this series was expanded in 1994 to include crimes motivated by victims' physical or mental disability.

- *Supplementary homicide reports:* Since 1962, reporting agencies have also been asked to complete Supplementary Homicide Reports (SHR).

---

[1]The 21 offenses currently tallied as Part II offenses are other assaults; forgery and counterfeiting; fraud; embezzlement; stolen property (buying, receiving, possessing); vandalism; weapons (carrying, possessing, etc.); prostitution and commercialized vice; sex offenses; drug abuse violations; gambling; offenses against the family and children; driving under the influence; liquor laws; drunkenness; disorderly conduct; vagrancy; all other offenses; suspicion; curfew and loitering laws (persons under 18); and runaways (persons under 18) (Federal Bureau of Investigation, 2004:8).

D–2   NATIONAL INCIDENT-BASED REPORTING SYSTEM

The *Uniform Crime Reporting Program Handbook* (Federal Bureau of Investigation, 2004:3) describes the origin of a new, more detailed format for the UCR as follows:

> By the 1980s, law enforcement was calling for a complete overhaul and modernization of the UCR Program. At a conference on the future of UCR, which was held in Elkridge, Maryland, in 1984, participants began developing a national data collection system that would gather information about each crime incident. By the end of the decade, the National Incident-Based Reporting System (NIBRS) was operational. NIBRS collects data on each incident and arrest within 22 offense categories made up of 46 specific crimes called Group A offenses. For each incident known to police within these categories, law enforcement collects administrative, offense, victim, property, offender, and arrestee information. In addition to the Group A offenses, there are 11 Group B offenses for which only arrest data are collected. The intent of NIBRS is to take advantage of available crime data maintained in modern law enforcement records systems. Providing considerably more detail, NIBRS yields richer and more meaningful data than those produced by the traditional summary UCR system. The conference attendees recommended that the implementation of national incident-based reporting proceed at a pace commensurate with the resources and limitations of contributing law enforcement agencies.

The NIBRS incident report is quite intricate and allows for great flexibility in the coding of individual events: spanning 46 offense categories, each incident report can include up to 10 offenses, 3 weapons, 10 relationships to victim, and 2 circumstance codes.

Although development of NIBRS began with a 1984 conference, a major impediment to the system's usefulness is that the "pace commensurate with the resources and limitations of contributing law enforcement agencies" envisioned in 1984 has turned out to be extremely slow. As of September 2007, the Justice Research and Statistics Association[2] estimated that only about 25 percent of the nation's population is included in NIBRS-compliant jurisdictions. In all, about 26 percent of agencies that supply data to the UCR do so using the NIBRS format. Among the states that have not yet implemented NIBRS are California, New York, and Pennsylvania; in Illinois, the only NIBRS participant to data is the Rockford Police Department. Five states—Alaska, Florida, Georgia, Nevada, and Wyoming—have not yet specified any formal plan for participation in NIBRS.

---

[2]See http://www.jrsa.org/ibrrc/background-status/nibrs_states.shtml [12/1/07].

– E –

Other Victimization Surveys: International and U.S. State and Local Experience

Walker (2006) summarizes the basic design features of 62 victimization-related surveys conducted by various countries, as part of an inventory sponsored by the United Nations Economic Commission for Europe and the United Nations Office on Drugs and Crime. About two-thirds of these are standalone surveys specifically focusing on victimization, and the remainder are crime and victimization components of more general, omnibus surveys. All told, these surveys ranged in size from about 400 to 60,000 households (or up to 75,000 people); 7 countries' surveys include 10,000 or more persons *and* households, although several of these are omnibus population surveys that include a victimization component. The survey design features of selected international victimization surveys—including the National Crime Victimization Survey (NCVS)—are summarized in Table E-1. Farrington et al. (2004) provide descriptions of both official-report and survey data sources in several countries, and Lynch (2006) summarizes international victimization surveys as a source for cross-national comparison.

Some countries' victimization surveys clearly use the well-established U.S. NCVS as a template; so, too, do several of the victimization surveys that have been fielded by individual states in the United States. Due to their cost, state victimization surveys tend to be one-shot efforts, although some states (e.g., Minnesota) have conducted several replications.

Table E-1 Design Features of Nation-Specific Victimization Surveys

Design Feature	USA[a]	Australia[b]	Canada[c]	England & Wales[d]	Sweden[e]	Netherlands[f]	Scotland[g]	Switzerland[h]	Ireland[i]
Sample									
Cross-Section	No	Yes	Yes (rotated)	Yes (rotated)	Yes	Yes	Yes (rotated)	Yes	Yes
RDD	No	No	Yes	No	No	No	No	No	No
Clustered							Yes		Yes
Screening									
Unified	No	Yes	Yes	No	Yes	Yes	No	Yes	Yes
Dense cuing	Yes	No	No	No	No		No	No	No
Bounding	Prior Interview	No	No	No	No	No	No	No	No
Reference period	6 mo.	12 mo.	12 mo.	12 mo.	12 mo.	12 mo.	12 mo.	Short/long ref period 12 mo.	12 mo.
Mode									
In-person	Yes	No	No	Yes	Yes	No	Yes	No	No
Phone	Yes	No	Yes	No	No	No	No	yes	No
Central CATI	Yes[j]	No	Yes	No	No	No	No	Yes	No
Self-administered	No	Yes	No	Yes	No	Yes	No	No	No
Series	Yes	No	No	Yes	Yes	No	Yes	Yes	Yes
Free-standing survey	Yes	No	No	Yes	No	Yes	Yes	Yes	No

[a] National Crime Victimization Survey.
[b] Summarizes three national crime victim surveys; see Carcase (2004).
[c] General Social Survey–Victimization.
[d] British Crime Survey.
[e] Victimization component of annual Survey of Living Conditions.
[f] Victim survey conducted by Statistics Netherlands; other victim surveys are conducted by the police, as described by Bijleveld and Smit (2004).
[g] Scottish Crime Survey; for 2004, the survey was renamed the Scottish Crime and Victimisation Survey.
[h] Swiss crime survey, which was used as a basis for the International Crime Victimization Survey; see Killias et al. (2004).
[i] Crime and Victimization component of Quarterly National Household Survey; see Central Statistics Office (2007).
[j] Calls from Census Bureau CATI centers to be eliminated from NCVS in 2007.

NOTES: RDD, random digit dialing; CATI, computer-assisted telephone interviewing.

SOURCE: Adapted from Lynch (2006:Table 2) and Farrington et al. (2004).

In this appendix, we provide additional detail on selected international victimization surveys that are particularly relevant to our discussion of NCVS design features in the main body of the report. We describe the British Crime Survey and the International Crime Victimization Survey in particular. We also provide a fuller description of subnational victimization surveys that have been conducted in the United States. (One long-standing source of some crime victimization information at the state level—the Texas Crime Poll—is not described here, but is summarized in Box 4-2.)

E–1 BRITISH CRIME SURVEY

The British Crime Survey (BCS) measures criminal victimization among the population age 16 and older in England and Wales; Scotland was included in the earliest versions of the survey, but it now conducts its own victimization survey (as does Northern Ireland). The BCS became an annual data collection in 2001, having previously been conducted on a roughly biennial basis (1982, 1984, 1988, 1992, 1994, 1996, 1998, and 2000). The BCS was originally conceived as "a research tool" designed to "obtain a better count of crime," "identify risk factors in victimization," and "examine people's worry about crime and their perceptions of and contact with the police." Only with the passage of time did the goal of "provid[ing] more reliable information about trends in crime" become a principal focus of the survey (Jansson, 2007:4).

The BCS is administered by the Home Office of the United Kingdom, the duties of which correspond—in part—to those of the U.S. Department of Justice. Unlike the NCVS, for which the Census Bureau is the data collection agent, the United Kingdom's principal statistical agency and data collector—the Office for National Statistics—plays no role in the conduct of the BCS. (However, the placement of the BCS between the Home Office and the Office for National Statistics has been debated in reviews of the British government's crime statistics programs, discussed in Section E–1.c.) An external research organization, BMRB Social Research, has been engaged as the data collection agent for the BCS since 2001. However, the contract to conduct the BCS is retendered every three years (Walker, 2006:3–4).

Owing to its annual nature, the BCS employs a 12-month reference period; interviews are conducted continuously throughout the year, so that 1-year window shifts depending on the interview date. The BCS target population consists of "households in England and Wales living in private residential accommodation" and the "adults aged 16 and over living in such households"; accordingly, the BCS does not attempt to interview institutional populations (Grant et al., 2007:§2.2).

E–1.a   Sample Design

The BCS uses the British Postcode Address File as its frame; "whole post-code sectors" are the primary sampling units (PSUs) for the survey and are selected after stratifying by basic demographic variables and Police Force Area (PFA) (Grant et al., 2007:§2.4–2.6). The current design of the BCS includes a base sample as well as two specialized "boost" components.

As of 2004, the BCS base sample is designed to ensure representation—through at least 1,000 interviews—in each designated PFA in England and Wales. The Home Office's decision to make the BCS an annual survey beginning in 2001 was accompanied by a doubling of sample size, since the generation of BCS estimates at the PFA level was also taken as a goal for the survey (Smith, 2006:3). Hence, for 2005–2006, the total BCS base sample size was approximately 47,000 (Grant et al., 2007). Sampling targets are still constructed so that the larger jurisdictions are allocated higher numbers of interviews—e.g., the "Metropolitan" PFA consisting of the Metropolitan (London) Police and the City of London Police was targeted for about 3,500 interviews in 2005–2006. To promote some continuity, half of the PSUs used in one year's sample (e.g., 2004–2005) are retained for use in the next year's sample (2005–2006), although fresh addresses are selected from those PSUs (and lists are checked so that no address is selected two years in a row).

Over time, the BCS response rate has ranged from 73 to 83 percent; it has been at about 75 percent since 2001 (Jansson, 2007:5).

E–1.b   Supplements in the BCS

From 2003 to 2006, the Home Office conducted a conceptual comple-ment to the self-response victimization survey of the BCS: the Offending, Crime and Justice Survey (OCJS), which is a national self-report survey of *offending* behavior. (The survey also covers other antisocial behaviors, such as drug and alcohol use; questions on perceptions of antisocial behavior had earlier been fielded as a module in the 1992 survey.) The OCJS is particu-larly oriented at measuring juvenile delinquency and so relaxes the BCS age constraint, collecting information from respondents as young as 10 years old and oversampling persons ages 10–25 so that they represent roughly half of the basic sample. The OCJS is designed as a longitudinal study, with multi-ple contacts of the same people, in order to permit measures of trajectories of violence within the study period. The size of the core sample beginning in 2003 was just over 10,000 people; by 2006, about 4,100 of the respondents had been contacted in previous waves of the survey, and a fresh sample brought the total sample size to about 5,000. The OCJS featured multi-ple response modes within the same interview: standard computer-assisted personal interviewing (CAPI) was used for most questions, but computer-assisted self-interviewing (CASI)—in which respondents read questions from a laptop computer screen and answered them directly—was used for more sensitive topics. Audio-CASI, in which respondents could hear the questions through headphones, was used for some respondents with language difficulties.

Other modules or supplements that have been added to the BCS over time include (Jansson, 2007):

- Perceptions of and confidence in the criminal justice system, including judges and the probation service (conducted regularly since 1996);

- Interpersonal violence, using CASI methods to try to promote fuller reporting of sensitive incident types, such as domestic violence, sexual victimization, and stalking (first conducted in 1994 and regularly thereafter); and

- Separate modules of questions on mobile phone theft (first in 2001), fraud and technology crimes (first in 2002), and identify theft (first in 2005).

E–1.c Reviews of British Crime Statistics

As part of a larger review of official statistics collected in the United Kingdom and their effectiveness in meeting the needs of users, a Statistics Commission (2006) was established and issued a final report on crime statistics in July 2006. Separately, the Home Secretary tasked an independent review board headed by Smith (2006:1) as follows:

> The Home Secretary is concerned that public trust in the crime statistics produced by the Home Office has declined to such an extent that it is no longer possible to have a debate about alternative criminal justice policies on the basis of agreed facts about the trends in crime. He wishes to be advised on what changes could be made to the production and release of crime statistics so that public trust is re-established.
>
> In addition, he wants the Review Group to examine the key issues raised by the Statistics Commission about crime statistics and to make practical recommendations to the Home Secretary as to what changes are needed to address those issues. . . .

The Smith (2006) review was published in November 2006.

In their focus on user needs, the basic tasks of these two review efforts in the United Kingdom are similar to our panel's task, although their mandates are considerably broader. The conduct of the program for official reports to police was in scope for both studies (while administration of the Uniform Crime Reporting program is not part of ours), let alone the Statistics Commission's broader task of reviewing data collections on other social and

---

Box E-1 Recommendations of the United Kingdom Statistics Commission (2006)

1. Responsibility for the compilation and publication of crime statistics should be located at arm's length from Home Office policy functions and with clear accountability within the evolving framework of the government statistical service.

2. Treasury and Home Office Ministers should consider together a fully developed business case for moving responsibility for the British Crime Survey to the Office for National Statistics and should publish their agreed view with supporting arguments.

3. The Home Office, and others as appropriate, should make changes to the presentation of the recorded crime figures in order to communicate better the main messages. These steps include:

   • changing the definition of violent crime;

   • greater distinction between British Crime Survey results and police recorded crime data and the uses for which each source is appropriate;

   • ensuring regular reviews of statistical classifications.

4. Existing local data should be better used to improve the quality and range of statistics on crime. This could be achieved through police forces agreeing to publish, in a co-ordinated way, standardised comparable analyses at a local level. These analyses need not necessarily be drawn together and published as official statistics by the Home Office but must be consistent with those that are.

5. Comparability of crime statistics between the various countries within the UK should be improved, identifying and addressing areas of statistics where there are problems.

6. Technical research should be carried out (to a published timetable) to develop a set of weighted index measures of 'total crime' and promote debate on which, if any, of these measures should be adopted alongside the current basic count.

SOURCE: Excerpted from Statistics Commission (2006).

---

demographic characteristics. For reference, the six formal recommendations of the Statistics Commission are listed in Box E-1, and selected recommendations (those most relevant to the BCS and the topics in this report) of the Smith commission are listed in Box E-2.

The Statistics Commission (2006:7) expressed particular concern "that the broad statistical messages about crime" from both the BCS and the official police report data "were being lost against a backdrop of confused reporting." Hence, its Recommendation 3 stresses the need for statistical communications to be clear about the nature and limitations of their data sources. The Smith independent review noted the value of complementary information provided by the BCS and official police reports, recommending that results from the two series continue to be published together (2.7) and that main messages from *both* series (and others, as appropriate) be explored

Box E-2 Selected Recommendations of the Smith (2006) Independent Review of British Crime Statistics

2.1 The British Crime Survey sample frame should be extended to include those under 16 and those living in group residences as soon as practical after taking the advice of those with relevant expertise and piloting the changes. In addition, research should be carried out on the victimisation of homeless and institutionalised populations.

2.2 We recommend that the Home Office should carry out a survey of commercial and industrial victimisation every two years.

2.3 We recommend that the Home Office should publish within 12 months an action plan for what it proposes to do to measure those crimes which are either not included in the present crime statistics or are poorly measured by them.

2.4 We recommend that the Home Office set up a standing panel of independent experts to provide regular review of and comment on methodological and analytic issues relating to the BCS and its other crime surveys.

2.7 The Home Office should continue to publish police recorded crime data and the BCS together.

2.8 We recommend that national crime statistics should be published annually and include a full commentary on the state of crime, drawing on all appropriate data sources.

2.10 We recommend that whenever Home Office statistical reports include interpretation or assumptions on the part of the authors these should be flagged frankly and openly on their first appearance in the report and the basis of those judgements should be referenced and made available.

2.11 We recommend that the Home Office should attach to each of its statistical series a statement clearly identifying the strengths and weaknesses of the particular series and the aspects about which professional judgements may need to be made.

2.12 In order to build trust, the Home Office should ensure that the release and statistical commentary on national crime statistics are quite clearly separated from political judgements or ministerial comments and should ensure the accuracy of any statements made about the statistics, whether in press releases or ministerial comments.

2.13 We recommend that the Home Office redefine violent crime in crime statistics to only include those crimes which actually cause physical injury or where the threat to inflict such injury is likely to frighten a reasonable person.

3.1 The Home Office should make the provision of local crime information a central part of its crime communication strategy and not just rely on publishing national crime statistics.

4.1 We recommend that the Home Secretary should put in place a regulatory environment which ensures that there is an actual and perceived separation between those who produce statistical data and commentary on crime (a "Back Office" function) and those who are responsible for policy advice and will be judged on the basis of the data (the "Front Office") be they in a police force, the Home Office or elsewhere.

SOURCE: Excerpted from Smith (2006).

by analysts (2.8), while taking care not to attribute findings or conclusions to the government or the Home Office generally (2.10).

Several of the commission's recommendations are driven by a concern about crime statistics being issued by the program agency responsible for developing policy on crime. That is, the commission "[did] not believe trust [in crime statistics] can be built up whilst the same Ministers, advisers and senior officials are directly involved both in publishing the figures and in setting out the Government's position" (Statistics Commission, 2006:12). The Smith (2006:2) review summarized the basic problem:

> Every Home Secretary [has] accepted that a major purpose of Home Office policy is to reduce crime and for the present government this has meant setting performance targets for crime reduction. As a result, crime statistics have become a key metric for judging the performance of the Home Office and therefore central to debates between government and opposition. This has meant that crime statistics have been subject to a quite new degree of scrutiny and their release and handling have become politically much more sensitive.

Hence, the commission's Recommendation 1 emphasizes the need for an arm's length distance between the Home Office's statistical operations and its policy operations. For the BCS, in particular, the commission's Recommendation 2 suggested a way for a more-than-arm's-length separation: transferring authority for the BCS from the Home Office to the Office for National Statistics (ONS), particularly if efforts to provide ONS greater independence as a nonministerial department came to fruition.[1] The Smith independent review reiterated the need for a clear divide between statistical releases and policy decisions (Recommendation 2.12) but disagreed on the administrative placement of the BCS. "We believe [the BCS] should remain in the Home Office because as well as being a source for the national crime statistics, it is one of the most important research tools and sources of information for the Home Office to manage the crime problem" (Smith, 2006:6).

The Smith review argued for expanding BCS coverage to persons under age 16 as well as to people in group housing and, possibly, institutions (Recommendation 2.1). Smith (2006:10) recognized that these changes "would be methodologically difficult" and would require careful development and piloting. The interviewing of children, in particular, raises parental consent issues that would take great care; the review text takes the more nuanced position that "interviews should be extended as far below 16 as proves prac-

---

[1] "Although we see the case for transferring responsibility as strong, the Commission does not have available to it all the relevant information on costs and capacity to make a firm recommendation"—hence the recommendation's call for a "business case" for transferring authority for the BCS (Statistics Commission, 2006:13).

tical in a regular household-based survey." Echoing the commercial victimization surveys that were originally part of the U.S. National Crime Survey—but dropped on the recommendation of the National Research Council (1976b)—the Smith review recommended that consideration be given to a biennial standalone study of commercial and industrial victimization (2.2).

Both the Statistical Commission and the Smith review observed the "substantial demand from local delivery organisations for more and better local level information on crime" (Statistics Commission, 2006:15).

E–2 INTERNATIONAL CRIME VICTIMIZATION SURVEY

Administered by the United Nations Interregional Crime and Justice Research Institute (UNICRI), the International Crime Victimization Survey (ICVS) is a set of surveys that aspires to provide standardized measures of victimization for cross-national comparison. A first round of the ICVS was fielded in 1989, after two years of development by a working group; subsequent rounds were fielded in 1992, 1996–1997, and 2000. The survey began with about 15 countries (and 2 separate cities), while the 1996–1997 round was said to interview 135,465 people in 56 countries.

The ICVS is conducted through computer-assisted telephone interviewing (CATI) supplemented with personal interviewing; the face-to-face personal interviews are typically restricted to a sample of people in a participant nation's capital city.[2] A survey firm from the Netherlands, Interview, coordinated the work in most of the participating industrialized countries, subcontracting with firms or companies in those countries for sample selection and fieldwork. ICVS participants bear the costs of fieldwork in their countries; they use a core set of questions in their interviews—pursuant to the goal of standardization—but have some flexibility for adapting the sample or the survey instrument based on their own needs. For instance, in Australia, the Australian Institute of Criminology oversaw the 7,000 ICVS interviews conducted in that country in the 2004 wave of the survey. The institute customized the ICVS to attend to matters of interest to the Australian government by making 1,000 of the interviews a "booster sample" of migrants from Vietnam and the Middle East, as well as by adding groups of questions on such issues as licensing and storage of firearms and perceptions of safety while using public transportation. They also dropped questions on sexual assault, judging that the institute's separate Violence Against Women Survey provided more reliable results (Johnson, 2005).

---

[2]Exceptions—cases in which personal interviews are favored over CATI—include places in which telephone penetration was not deemed to be high, including Malta, Northern Ireland, and rural Spain. ICVS work in Japan was also done face-to-face (http://www.unicri.it/wwd/analysis/icvs/methodology.php).

The ICVS covers three crimes against persons (assaults and threats, robbery, and personal theft) and six crimes against households (burglary, attempted burglary, motor vehicle theft, theft from motor vehicles, motorcycle theft, and bicycle theft). Due to the frequency of the survey, the ICVS uses a 5-year reference period, asking respondents to recall incidents within that range; however, it also generally asks about events that occurred within the last 12 months, so that 1-year estimates may also be produced.

As described on the ICVS methodology web page,[3] ICVS sample selection typically includes 1,000–2,000 households in participant households, reached through random-digit dialing (RDD) methods. A randomly selected person from each household is chosen to complete the interview (and is not replaced if that person refuses to participate). Contacts continue until the desired number of interviews is completed. Overall, this CATI component had a 67 percent response rate in 11 industrialized countries in the 1996 ICVS round, with rates varying from 40 percent (in the United States) to 80 percent or greater.

When face-to-face interviewing was used, procedures tended to be somewhat more ad hoc. Effort was typically concentrated in the country's capital city, with the intent of obtaining about 1,000 respondents. The ICVS methodology page indicates that the sampling for face-to-face interviews "was generally hierarchical," starting "with identifying administrative areas within the city, followed by a step-by-step procedure aiming at identifying areas, streets, blocks, households and finally the household member aged over 16 whose birthday is next."

E–3   VICTIMIZATION SURVEYS AS PART OF A BROADER SOCIAL SURVEY: CANADA AND AUSTRALIA

Some countries that do not have standalone surveys of victimization still gather related information through major supplements to other, more omnibus surveys. Two examples are Canada and Australia.

Canada's system for the measurement of crime and victimization is similar to that of the United States. Official counts of crimes reported to the police have been collected annually since 1962 by the Uniform Crime Reporting Surveys (e.g. Gannon, 2006), while victimization measures are obtained as part of Statistics Canada's omnibus General Social Survey.[4] Since 1988, a victimization component in the General Social Survey has been in-

---

[3] See http://www.unicri.it/wwd/analysis/icvs/methodology.php.

[4] Despite the "Surveys" part of the name, the Canadian UCR is intended as a complete census of Canadian law enforcement agencies. As a further parallel to the U.S. model, the Canadian Uniform Crime Reporting Surveys include a component dubbed UCR2—gathering detailed incident-based information—that is similar to the developing American National Incident-Based Reporting System (NIBRS; see Section D–2). As of 2005, 127 Canadian police depart-

cluded on a quinquennial basis (however, the third round was conducted in 1999, not 1998). Starting with the 1999 administration, the survey was renamed the General Social Survey–Victimization (GSS-V, replacing General Social Survey–Personal Risk).

The GSS-V is strictly a telephone survey, conducted by RDD from three centralized CATI centers in Statistics Canada regional offices. A Statistics Canada web page describing the survey's methodology notes that the RDD approach excludes persons in households without telephones (estimated as representing "less than 2% of the target population" of Canadians 15 years of age or older) and persons with only cellular telephone service ("again, this group makes up a very small proportion of the population, less than 3%").[5]

Consistent with the survey's frequency, the GSS-V asks respondents to recall incidents within the past five years. Working with this long reference period can require repeated sets of questions, since circumstances may have changed within the time window. For instance, the first set of questions related to physical or sexual abuse by a spouse or partner asks "whether, in the past 5 years, your *current* spouse/partner has done any of the following to you" (emphasis in original). A follow-up question asks whether such acts of violence have occurred more than one time; if so, respondents are asked how many times the abuse has taken place in 12 months. In either event, the specific month and year is sought for the most recent incident. Subsequently, another module of questions asks about incidents of abuse at the hands of one's previous spouse or partner.

Like the NCVS, the GSS-V has included topical modules sponsored by other agencies: specifically, the Interdepartmental Working Group on Family Violence funded questions on domestic and elder abuse, and the Solicitor General Canada funded questions on public perceptions on imprisonment in the 1999 administration of the survey. However, both of these sponsored supplements were dropped from the 2004 GSS-V.

The GSS-V sample was selected based on 27 strata, with major cities (census metropolitan areas) representing their own strata and partitioning nonmetropolitan areas into about 10 strata. Within each strata, respondents were contacted through RDD after eliminating "non-working banks" (telephone exchanges known not to work). When a household was contacted, basic demographic information on all household members was elicited; one household member age 16 or older was then randomly selected from this list to complete the interview.

---

ments completed the UCR2 survey, "represent[ing] 62% of the national volume of reported" crimes (Gannon, 2006:14).

[5] See http://www.statcan.ca/cgi-bin/imdb/p2SV.pl?Function=getSurvey&SDDS=4504&lang =en&db=IMDB&dbg=f&adm=8&dis=2; as of June 21, 2007, this page showed a "last modified" of October 27, 2005.

For the 2004 administration of the Canadian survey, the overall response rate was estimated as 75 percent.

The Australia Crime and Safety Survey—administered by the Australian Bureau of Statistics—is a periodic supplement to the omnibus Monthly Population Survey. Specifically, it is part of the April Labour Force Survey, itself a large supplement to the Monthly Population Survey. Most recently, the Crime and Safety Survey was conducted between April and July 2005; it had previously been fielded at the national level in 1975, 1983, 1993, 1998, and 2002 (Australian Bureau of Statistics, 2006).

E–4   STATE AND LOCAL VICTIMIZATION SURVEYS

Requests for proposals (RFPs) under the Bureau of Justice Statistics (BJS) State Justice Statistics Program for Statistical Analysis Centers (SACs) in 2000 identified local victimization surveys as a high priority. Specifically, the RFP indicated that "SACs receiving funds under this theme must agree to use the BJS developed *Crime Victimization Survey* software, which can be easily modified to meet State/local priorities and requirements" (Alaska Justice Statistical Analysis Center, 2002:1).

E–4.a   Alaska

The Alaska SAC—the Justice Center at the University of Alaska, Anchorage—secured BJS funding in fall 2000 to conduct a localized version of the NCVS, with the intent of providing victimization estimates for the city of Anchorage.

At the time of the original contract award, a planning group was to consider methods for conducting a victimization survey for rural Alaska; formal work in that direction does not appear to have occurred. More recently, the Alaska SAC has switched focus from a dedicated victimization survey to a general Alaska Community Survey.

E–4.b   Illinois

In 2002, the Illinois Criminal Justice Information Authority contracted with Bronner Group, LLC, to conduct its first victimization survey. The survey was sent by mail to about 7,500 residents; the sample was drawn by the Illinois secretary of state's office from its database of holders of driver's licenses and state-issued identification cards. Unlike Minnesota's victimization survey, described below, the sample was apparently filtered to include only persons age 18 or older. The survey methodology and results are documented by Hiselman et al. (2005); of the initial sample, some 23 percent proved to be unreachable (e.g., the person had moved or was deceased). In all, 1,602 completed questionnaires were received, for a response rate of 28 percent (Hiselman et al., 2005:v).

E–4.c   Maine

The first Maine Crime Victimization Survey was conducted in fall 2006 by the University of Southern Maine's Muskie School of Public Service, which houses the state's SAC. A telephone questionnaire was administered to 803 adults (RDD until target sample size reached) between August and December 2006. The survey report (Rubin, 2007) indicates that the effort in Maine was modeled after Utah's victimization survey (Haddon and Christenson, 2005), described below. The Maine survey also adopted questions on identity theft from the 2004 NCVS.

E–4.d   Minnesota

The Minnesota Crime Survey is one of the few state victimization surveys to be fielded multiple times on a semiregular basis (1993, 1996, 1999, 2002, and 2005). This three-year cycle has allowed the content of the survey to evolve over time; for instance, the 2002 survey added a question about perceived fear of a terrorist attack as well as additional questions on domestic abuse (Minnesota Justice Statistics Center, 2003). Conducted by Minnesota Planning, the survey is conducted by mail using driver's license (or state-issued identification card) rolls as the sampling frame. Generally, the use of this frame means that the survey covers persons age 16 and older, since that is the legal age for obtaining a driver's license; however, since identification cards can be issued to persons of any age, it is possible for someone under 16 to be selected for the survey.

The Minnesota survey used an advance-notice postcard to alert sampled respondents that a questionnaire should soon arrive, and a reminder postcard was issued if no response was received within three weeks. Survey collection ended after five weeks. In 2002, the survey obtained a 41.6 percent response rate on a mailout of 10,013 questionnaires.

E–4.e   Utah

Utah's BJS-affiliated SAC is the Research and Data Unit of the state Commission on Criminal and Juvenile Justice, in the office of the governor. In 2005, the Utah SAC conducted its third Utah Crime Victimization Survey; it used a 12-month reference period, so that respondents were asked to report crimes that occurred in calendar year 2004. However, the 2005 version of the survey also added a "lifetime victimization" question to each of the specific crime types covered by the survey. The 2005 survey also expanded a battery of questions about attitudes and perceptions of crime (e.g., "When you leave your home, how often do you think about it being broken into or vandalized?").

Unlike its predecessors—covering crimes in 2000 and 2002—the 2005 administration of the Utah victimization survey was the first to be conducted by telephone. Households were reached by RDD, with calls made until a target sample size was achieved. The 2005 administration of the survey included 2,002 interviews and found that 41.3 percent of the respondents had been a victim of one of the crimes covered by the survey in 2004, a slight increase from 2002's estimate of 36.6 percent (Haddon and Christenson, 2005:4).

The Utah SAC conducted a separate study—focused on women age 18 and older—specific to rape and sexual violence. Comparing this survey to the state victimization survey, Haddon and Christenson (2005:18–19) note that the separate study's findings about the likelihood of a respondent's having been raped sometime during her lifetime were "not unlike those of the 2004 victimization survey"; however, the standalone survey suggested a much lower level of reporting to the police. They conclude that the lower reporting total "is likely more accurate in that [the rape survey] included a much larger group of individuals who had been sexually victimized."

E–4.f Wyoming

With support from BJS, the Wyoming Statistical Analysis Center (2004) and the Wyoming Division of Victim Services fielded a victimization survey in October 2003. The survey was conducted by random-digit dialing with an extension list stratified by county with selection probability proportional to population size. The survey yielded 1,439 interviews.

Anticipating a relatively low victimization rate, the survey was structured so that interviews did not simply end if a respondent indicated no incidents in the screener questions. Instead, those respondents were routed through a set of questions on attitudes toward criminal justice and awareness of the Division of Victim Services, and those respondents who did experience victimizations were guided through the analog of the NCVS incident form.

– F –

Biographical Sketches of Panel Members and Staff

Robert M. Groves *(Chair)* is professor of sociology and the director of the Survey Research Center in the Institute for Social Research at the University of Michigan. He is the author of *Survey Errors and Survey Costs* and the coauthor of *Nonresponse in Household Surveys*. A National Associate of the National Academies, he has served on seven National Research Council (NRC) committees and is a former member of the Committee on National Statistics. From 1990 to 1992, he served as associate director for statistical design, standards, and methodology at the U.S. Census Bureau. He is a fellow of the American Statistical Association and an elected member of the International Statistical Institute, and he has received the Innovator Award and an award for exceptionally distinguished achievement from the American Association for Public Opinion Research. He has an M.A. in statistics, an M.A. in sociology, and a Ph.D. in sociology, all from the University of Michigan.

William G. Barron, Jr., is a consultant to Princeton University and currently serves as consultant to the deputy director of the U.S. Census Bureau. After a 30-year career at the Bureau of Labor Statistics—serving as deputy commissioner (1983–1988) and acting commissioner (once for a 23-month period)—he moved to the U.S. Census Bureau in 1998. There he served as deputy director and chief operating officer. Heavily involved in the conduct and completion of the 2000 census and the development of plans for the 2010 census, he served as acting director of the Census Bureau in 2001 and early 2002. Prior to his consultancy at Princeton, he was visiting lecturer and Frederick H. Shultz Class of 1951 professor of international economic policy, and later the John L. Weinberg/Goldman Sachs and Company visiting professor and lecturer at the university's Woodrow Wilson School of Public and International Affairs. He has served as senior vice president for economic studies at the National Opinion Research Center (NORC) at the University of Chicago and as senior client executive at Northrop Grumman Corporation. He has a B.A. from the University of Maryland.

William Clements currently serves as dean of the School of Graduate Studies and professor of criminal justice at Norwich University. Prior to assuming the role of dean, he was director and creator of the Master of Justice Administration program (2002–2005) and executive director of the Vermont Center for Justice Research (1994–2005), Vermont's Bureau of Justice Statistics-affiliated Statistical Analysis Center. He has been involved in bringing Norwich's curriculum to the online environment and developing the online graduate program model. His professional research interests and experience include a variety of criminal justice system studies in program evaluation, data systems development, and adjudication patterns. He was most recently appointed by the Vermont Supreme Court as vice-chair of the newly formed Vermont Sentencing Commission and has worked on and published in the areas of incident-based crime data, juvenile justice, the operation of the courts, and sentencing trends. He has served in various capacities and as president of the Northeast Academy of Criminal Justice Sciences, and he is a past president and executive committee member of the Justice Research and Statistics Association (JRSA). He is coeditor of *Justice Research and Policy*. He has a Ph.D. in sociology from the University of Delaware.

Daniel L. Cork *(Study Director)* is a senior program officer for the Committee on National Statistics, currently serving as study director of the Panel to Review the Programs of the Bureau of Justice Statistics and costudy director of the Panel on the Design of the 2010 Census Program of Evaluations and Experiments. He previously served as study director of the Panel on Residence Rules in the Decennial Census, costudy director of the Panel on Research on Future Census Methods, and program officer for the Panel to Review the 2000 Census. His research interests include quantitative criminology, particularly space-time dynamics in homicide; Bayesian statistics; and statistics in sports. He has a B.S. in statistics from George Washington University and an M.S. in statistics and a joint Ph.D. in statistics and public policy from Carnegie Mellon University.

Janet L. Lauritsen is professor of criminology and criminal justice at the University of Missouri–St Louis. Much of her research is focused on understanding individual, family, and neighborhood sources of violent victimization as well as race and ethnic differences in violence. She served as chairperson of the American Statistical Association Committee on Law and Justice Statistics from 2004–2006 and as visiting research fellow at the Bureau of Justice Statistics from 2002 to 2006. During her fellowship, she assembled two expert meetings on major options for the National Crime Victimization Survey, several of the participants of which are also members of this panel. She currently serves on the editorial boards of *Criminology* and the *Journal of Quantitative Criminology* and on the executive board of the American Society of Criminology. She has a Ph.D. in sociology from the University of Illinois at Urbana-Champaign.

Colin Loftin is co-director of the Violence Research Group, a research collaboration with colleagues at the University at Albany and the University of Maryland that conducts research on the causes and consequences of interpersonal violence. The major themes of the research are (1) understanding violence as a social process extending beyond individual action, (2) improving the quality of data on the incidence and nature of crime, (3) the design and evaluation of violence prevention policies, and (4) the investigation of population risk factors for violence. The Violence Research Group published the *Statistical Handbook on Violence in America*. A past member of the National Research Council's Committee on Law and Justice, he previously served on the Panel on Understanding and Preventing Violence. He has a Ph.D. in sociology from the University of North Carolina.

James P. Lynch is distinguished professor at John Jay College of Criminal Justice in New York. At the Bureau of Social Science Research in the 1980s, he served as manager of the National Crime Survey redesign effort for the bureau. He became a faculty member in the Department of Justice, Law and Society at American University in 1986, where he remained as associate professor, full professor, and chair of the department until leaving for John Jay in 2005. He has published 3 books, 25 refereed articles, and over 40 book chapters and other publications. He was elected to the executive board of the American Society of Criminology in 2002 and has served on the editorial boards of *Criminology* and the *Journal of Quantitative Criminology* and as deputy editor of *Justice Quarterly*. He has also chaired the American Statistical Association's Committee on Law and Justice Statistics. He has a Ph.D. in sociology from the University of Chicago.

Ruth D. Peterson is professor of sociology and director of the Criminal Justice Research Center at Ohio State University, where she has been on the faculty since 1985. She is also a fellow of the National Consortium of Violence Research, where she coordinates the Race and Ethnicity Research Working Group. She has conducted research on legal decision making and sentencing, crime and deterrence, and most recently, patterns of urban crime. She is widely published in the areas of capital punishment, race, gender, and socioeconomic disadvantage. Her current research focuses on the linkages among racial residential segregation, concentrated social disadvantage and race-specific crime, and the social context of prosecutorial and court decisions. She has a Ph.D. in sociology from the University of Wisconsin.

Carol V. Petrie *(Senior Program Officer)* is director of the Committee on Law and Justice at the National Academies. She also served as the director of planning and management at the National Institute of Justice, U.S. Department of Justice, responsible for policy and administration. In 1994, she served as the acting director of the National Institute of Justice. She has conducted research on violence and public policy, and managed numerous research projects on the development of criminal behavior, domestic violence, child abuse and neglect, and improving the operations of the criminal justice system. She has a B.S. in education from Kent State University.

Trevillore Raghunathan is professor of biostatistics and research professor at the Institute for Social Research at the University of Michigan. He also teaches in the Joint Program in Survey Methodology at the University of Maryland. He is the director of the Biostatistics Collaborative and Methodology Research Core, a research unit designed to foster collaborative and methodological research with the researchers in other departments in the School of Public Health and other allied schools. He is an associate director of the Center for Research on Ethnicity, Culture and Health and a faculty member of the Center of Social Epidemiology and Population Health; he is also affiliated with the University of Michigan Transportation Research Institute. Before joining the University of Michigan in 1994, he was on the faculty in the Department of Biostatistics at the University of Washington. His research interests are in the analysis of incomplete data, multiple imputation, Bayesian methods, design and analysis of sample surveys, small-area estimation, confidentiality and disclosure limitation, longitudinal data analysis, and statistical methods for epidemiology. He has a Ph.D. in statistics from Harvard University.

Steven R. Schlesinger is chief of the Statistics Division at the Administrative Office of the United States Courts. He was director of the Bureau of Justice

Statistics from 1983 to 1988 and was deputy director of the U.S. Department of Justice Office of Policy and Communications from 1991 to 1993. He has also taught on the political science faculties of Rutgers University and the Catholic University of America. He is the author of 2 books and over 25 articles on legal topics. Among his professional awards are the O.J. Hawkins Award for Innovative Leadership and Outstanding Contributions to Criminal Justice Systems, Policy and Statistics in the United States, the U.S. Attorney General's Award for Excellence in Management, and AO's Meritorious Service Award. He has a Ph.D from the Claremont Graduate School.

Wesley G. Skogan has been a faculty member at Northwestern University since 1971 and holds joint appointments with the political science department and the University's Institute for Policy Research. His research focuses on the interface between the public and the legal system, crime prevention, victim services, and community-oriented policing. He has written four books on policing; all are empirical studies of community policing initiatives in Chicago and elsewhere. His 1990 book *Disorder and Decline* examined public involvement in these programs, their efficacy, and the issues involved in police-citizen cooperation in order maintenance. Another line of his research concerns neighborhood and community responses to crime. He has edited a series of technical monographs on victimization research and authored a technical review of the National Crime Victimization Survey that was published in *Public Opinion Quarterly*. He served as a consultant to the United Kingdom Home Office, developing and analyzing the British Crime Survey. He has been a visiting scholar at the Max-Planck-Institut (Freiburg), the Dutch Ministry of Justice (WODC), the University of Alberta, and Johns Hopkins University. He spent 2 years as a visiting fellow at the National Institute of Justice. At the NRC, he has served on the Committee on Law and Justice and chaired the Committee on Research on Police Policies and Practices. He has a B.A. in government from Indiana University, an M.A. in political science from the University of Wisconsin, and a Ph.D. in political science from Northwestern University.

Bruce D. Spencer is professor of statistics and faculty fellow in the Institute for Policy Research at Northwestern University. His interests include the interactions between statistics and policy, demographic statistics, and sampling. He chaired the statistics department at Northwestern from 1988 to 1999 and 2000 to 2001. He directed the Methodology Research Center of the National Opinion Research Center (NORC) at the University of Chicago from 1985 to 1992. From 1992 to 1994 he was a senior research statistician at NORC. At the National Research Council he served as a member of the Panel on Formula Allocations of the Committee on National Statistics and the Mathematical Sciences Assessment Panel and the Panel on Statistical

Issues in AIDS Research; as a staff member he served as study director for the Panel on Small Area Estimates of Population and Income. He has a Ph.D. from Yale University.

Bruce Western is professor of sociology at Harvard University and director of the Multidisciplinary Program in Inequality and Social Policy at the Kennedy School of Government. Previously, he was professor of sociology at Princeton University and faculty associate in the Office of Population Research. His research interests broadly include political and comparative sociology, stratification and inequality, and methodology. More specifically, he has studied how institutions shape labor market outcomes. Work in this area has developed along two tracks: the growth and decline of labor unions and their economic effects in the United States and Europe; and the impact of the American penal system on labor market inequality. His methodological work has focused on the application of Bayesian statistics to research problems in sociology. He is the author of *Punishment and Inequality in America* and, with Mary Patillo and David Weiman, of *Imprisoning America: The Social Effects of Mass Incarceration*, both publications of the Russell Sage Foundation. He has a Ph.D. in sociology from the University of California, Los Angeles.